GOING NUCLEAR
IRELAND, BRITAIN AND THE CAMPAIGN TO CLOSE SELLAFIELD

For Margaret, beloved sister and friend. Long life, health and happiness.

GOING NUCLEAR
Ireland, Britain and the Campaign to Close Sellafield

VERONICA McDERMOTT

IRISH ACADEMIC PRESS
DUBLIN • PORTLAND, OR

First published in 2008 by Irish Academic Press

44 Northumberland Road,
Ballsbridge,
Dublin 4, Ireland

920 NE 58th Avenue, Suite 300
Portland, Oregon,
97213-3786

www.iap.ie

Copyright © 2008 Veronica McDermott

British Library Cataloguing in Publication Data
An entry can be found on request

ISBN 978 0 7165 2908 8 (cloth)
ISBN 978 0 7165 2909 5 (paper)

Library of Congress Cataloging-in-Publication Data
An entry can be found on request

All rights reserved. Without limiting the rights under copyright reserved alone, no part of this publication may be reproduced, stored in or introduced into a retrieval system, or transmitted, in any form or by any means (electronic, mechanical, photocopying, recording or otherwise), without the prior written permission of both the copyright owner and the above publisher of this book.

Typeset by Carrigboy Typesetting Services
Printed by Biddles Ltd., King's Lynn, Norfolk

Contents

Acknowledgements	vii
List of Illustrations	x
Introduction	xi

PART ONE: DISCOVERIES

1.	New Beginnings	3
2.	Splitting the Atom	15
3.	Darkening Clouds	24

PART TWO: THE BOMB MAKERS

4.	Weapon of Mass Destruction	35
5.	'Fat Man' and 'Little Boy'	44

PART THREE: NUCLEAR PROLIFERATION

6.	Britain's Bomb Factory	59
7.	Proliferation	73
8.	Heat and Light	86
9.	The Windscale Fire, 1957	100

PART FOUR: THE RISE AND FALL OF CIVIL NUCLEAR POWER

10.	Ireland's Opportunity	115
11.	The Plutonium Economy	128
12.	Windscale to Sellafield	140
13.	An Acceptable Risk	149
14.	The Legacy of Three Mile Island	156

PART FIVE: IRELAND V. BRITAIN: THE SELLAFIELD PHONEY WAR

15.	Atomic Village	171
16.	Close Down Sellafield	185
17.	Battle Stations	197
18.	Thorp	210
19.	A Political Bone of Contention	223
20.	Friends and Neighbours	238
21.	Pyrrhic Victories	253
22.	Conclusions	272

Notes	286
Select Bibliography	313
Index	317

Acknowledgements

I owe huge gratitude to Professor Richard Wakeford and Professor Ian McAulay, who, before a word of this book was committed to paper, agreed to review it for accuracy, and, true to their word, did that and a lot more. Also my friends, Paddy O'Shaughnessy, whose detailed knowledge of British and Irish history and international affairs was always there to point me in the right direction, and Jackie Gallagher, whose advice and practical help sustained my efforts and morale at crucial points in the writing of this book.

I am grateful to so many people who gave their personal time to share their experiences of the nuclear industry and their personal knowledge of Ireland's political dispute with Britain over Sellafield; who provided suggestions on material sources and helped with useful introductions. My thanks to all who agreed to be interviewed, especially Des O'Malley, whose recollections of Ireland's energy crisis in the 1970s and the controversy over the Carnsore Point nuclear project illuminated that period of Irish history. Also my thanks to Deputies Martin Mansergh and Fergus O'Dowd and Professor William Reville for their time and assistance, and to Dr Garret FitzGerald, who agreed to talk with me on the telephone.

A very special thanks is reserved for RPII Board member Frank Turvey, who not only gave of his time, energy and encouragement but opened many doors to other sources for interview, profitable lines of research and practical assistance. Also to Professor Philip Walton, for a wide-ranging interview on the life and times and notable achievements of his father, Ernest, Ireland's only scientific laureate.

A special debt of gratitude is owed to Jim Innes, who persistently reminded me that this opportunity to tell the story of the Sellafield workforce and their community in West Cumbria should not be missed. I hope I have done the workers of that community some justice in this account. My thanks are also due to Brian Wilson, former UK Energy Minister, who agreed to meet me, at short notice, on a wet morning in Glasgow and whose advice has been invaluable throughout.

I am grateful to Colin Duncan and his wife Elaine for their gracious hospitality whilst I was researching this project in the UK. Also my friend of many years, Gavin Carter, in Washington, for keeping me focused at all stages of this project, challenging my ideas to very good effect. To Tom McLaughlan, not least for almost missing his commuter train in London because he was reading drafts of the early manuscript; and to all others with whom I spoke over the past two years of research and writing, my heartfelt thanks. John Bowler of Greenpeace, who gave up a long Saturday afternoon to be interviewed for this book, is deserving of special mention, not alone for his frankness about his experiences at the helm of Greenpeace Ireland, but for his lifelong and inspiring commitment to the protection of nature and the environment.

The professional assistance of my publisher, Irish Academic Press, especially Lisa Hyde, is acknowledged. I will always be grateful for her unstinting support for this book. Thanks also to Andy Oppenheimer, who read some of the early draft chapters.

A special thanks to my son, Fergal Malone, for researching the newspapers of the earlier period and his discovery of the Glentoran Northern Ireland story. Also to my daughter, Nessa Malone, whose advice on writing popular history helped avoid many pitfalls. To Emer Brady, researcher and personal assistant throughout the summer of 2006, who undertook the tedious task of transcribing taped interviews and spent many other long days in the National Library and the Public Records Office in Dublin working on my behalf. The staff of all institutions and libraries visited for research purposes are acknowledged for their practical assistance and co-operation in locating and retrieving relevant source materials.

I am grateful also to all those who provided the photos for use in this book, especially Ali McKibbin, Press Office Manager at Sellafield, and Diane Gregg of British Nuclear Group Sellafield, who responded so promptly and efficiently to my request for assistance. Also Neil McCann, who at no little inconvenience to himself ensured that I received pictures of the Irish Sea Nuclear Free Flotilla.

Although Rupert Wilcox Baker was not formally interviewed for this book, the fact is that it could never have happened without him. In the ten years, on and off, that we worked together on BNFL's public affairs communications in Ireland, whatever else was going on around him, Rupert's personal integrity and goodwill towards Ireland was never in question.

Finally, there are no words adequate to express the depth of my gratitude for the patience, self-sacrifice, moral and practical support, tolerance, inspiration and boundless sense of humour of my husband, Tony Malone, who makes possible everything that's any good in this world.

List of Illustrations

1. Ernest and Freda Walton on their wedding day.
2. E.T.S. Walton as a small child with his sister Dorrie.
3. Windscale site in the early 1950s, dominated by the Windscale piles and tall chimneys.
4. Aerial view of Sellafield in 2005.
5. Lord Glentoran's enthusiasm for a nuclear plant in Northern Ireland was not universally shared in Whitehall.
6. Windscale under construction in 1950.
7. Construction workers install steelwork for radiation filters on the Windscale chimney stacks in March 1950.
8. The partly built Calder Hall reactor and cooling towers in the early 1950s.
9. The Windscale AGR, with its distinctive golf-ball shape, was the prototype for Britain's second generation nuclear power plants and an iconic landmark of the Sellafield site.
10. The decommissioning of 5,702m of old pipelines used to discharge radiaoactive wastes and rainwater from the Sellafield site to the Irish Sea from the early 1950s was finally completed in June 2006.
11. Des O'Malley, Minister for Energy, Industry and Commerce in the late 1970s, at the opening of a new factory in Ireland.
12. Ray Burke (left), long a central figure in Ireland's anti-Sellafield campaigns, pictured with Joe Jacob in 1997.
13. NCNI shop stewards Howard Rooms, John Kane and Douggie McCartney with their local MP, Jack Cunningham (second left), on their way to the first of several meetings with Prime Minister Tony Blair in Downing Street to plead the Sellafield cause.
14. The Sellafield workers' 'Trust Us' campaign was underwhelmed by Bono's attempts to stop Thorp.
15. Inside the Sellafield MOX plant.
16. The Irish yatch *Spinner*, and Greenpeace inflatables, confront the nuclear cargo ship the *Pacific Pintail*, at the approch to the harbour in Barrow-in-Furness in September 2002.
17. Irish Green Party MEP, Nuala Ahern (right foregound, holding banner) took a leading part in the Irish Sea flotilla protest.

18. Greenpeace ship *Rainbow Warrior* leads the protest of Irish yachts across the Irish Sea.
19. Irish yacht makes the message clear.
20. Cartoonist Martyn Turner's take on Sellafield's defence against terrorist attacks.
21. Labour Party Leader, Ruairi Quinn, in 2001 with party colleagues Emmet Stagg and Mary Upton take their skull and crossbones protest to the British Embassy in Merrion Road.
22. The Green Party's Trevor Sargent, with colleagues Ciaran Cuffe (left) and Eamon Ryan, planted paper windmills on the beach at the Poolbeg power station in Dublin to commemorate the twentieth anniversary of Chernobyl.

Introduction

At the heart of this book is the story of the development of nuclear power in Britain and Ireland and the infamous dispute that later developed between the two countries over Sellafield, making it a defining issue in Anglo-Irish relations in the last quarter of the twentieth century.

This British–Irish dispute serves as an allegory for the politics, public fears, logic, and sometimes its absence, that characterised the debate on nuclear energy that has raged passionately in Western society at various points for more than fifty years. Arguments have focused variously on the environmental and economic sustainability of nuclear power, fears of the long-term effects of radiation on human health, the problems of nuclear waste, and the morality of using a technology so intimately linked to weapons of mass destruction and the risks of proliferation. The development of military and civil nuclear power, as viewed through the prism of Ireland's own nuclear history and its relationship with Britain, provides valuable insights relevant to any critical analysis of the current international political environment and the challenges that confront us.

In these early years of the twenty-first century, humanity faces new challenges. Not least of these is global warming and how best we can respond to it, in so far as scientists can reasonably predict what its effects may be on our environment and human society throughout this century and beyond. Whatever the arguments about the contribution of economic and industrial activity to the build up of greenhouse gases believed to be fuelling potentially disastrous climate change throughout the world, there is no doubt that it is Western countries that bear primary responsibility for creating the problem and ultimately for fixing it. Ironically, it seems that the poorest countries and peoples with the least resources to adapt to climate change will be those on whom its impact will be most severe.

On the positive side, more than any other international issue of recent times, climate change underlines the interdependence of all human beings, irrespective of race, country or culture, forcing a re-evaluation of many of our most cherished prejudices about one another, whether these are grounded in political nationalism, doctrinal ideologies or technological bias.

The downside is also all too obvious: already there are signs of the emergence, especially among some Western media and politicians, of a new cult of eco-fundamentalism, whose zealots may seek to drive unwarranted and unworkable solutions to climate change based on irrational fears and unreasonable assumptions and oppress any voices dissenting from their doctrine of imminent and total catastrophe. On the opposite side lurk the climate change deniers, who for reasons of narrow self-interest would equally distort rational debate and seek to prevent appropriate corrective action.

As the experience of the nuclear debate over the past half-century can teach us, an extremist, politically driven, approach does nobody any favours in the long term. Such an approach to environmental issues, whether to nuclear power or global warming, is generally not underpinned by the observations of science. Rather, it relies more on rhetoric than on substance and essentially reflects the flawed mindset and specific political agenda of the doomsayers.

Understanding the origins of our particular prejudices, acknowledging and appreciating those origins in the context of their own time and place in history, is essential to making informed and intelligent decisions about the future. That is the core idea which has inspired this book.

* * *

October 2007 marks the fiftieth anniversary of the Windscale fire. Until Chernobyl in 1986, Windscale was the world's most serious nuclear reactor accident. The Windscale reactors, or 'piles', as they were called, were a military installation whose primary purpose was the production of plutonium and related materials for Britain's atomic bomb programme. Located at Sellafield in West Cumbria, their tall chimney stacks dominated the local landscape of rolling hills and sheep farms, set between the impressive mountain fells and the seashore. Metaphorically, they also cast a long shadow across the Irish Sea to Ireland.

Viewed from a distance, Ireland's twenty-five-year battle to close Sellafield and its emergence as a campaigning anti-nuclear state appears heroic. It has many universal facets, not least the battle of a small country against a big neighbour, an environmentally innocent David seeking to bring down a nuclear Goliath. Up close, the real picture reveals itself as infinitely more complex, comprising several distinct strands.

Among these is Sellafield as a feint in Ireland's battle to establish true independence from Britain throughout most of the twentieth century. A political perception of Sellafield as a 'safe issue' that posed no threat to Ireland's overall relationship with Britain, nor Ireland's own self-interests; and its sometimes flagrant manipulation for reasons of domestic political gain. A post-war Britain struggling to maintain its great power status and copper-fasten its 'special relationship' with the United States by acquiring an independent nuclear capacity. A frustrated Northern Ireland of the 1950s, clutching to its own dreams of a nuclear-fuelled future. The trials of a small community in West Cumbria in England's north-west striving to ensure their economic survival. A British state-owned enterprise, British Nuclear Fuels Limited (BNFL), entertaining grandiose Faustian ambitions of becoming an international world-class player in the nuclear industry.

Uncovering these stories begins with a journey into the past, tracing the exciting discoveries in atomic physics by a global community of independent scientists in the late nineteenth and early twentieth centuries who probed the secrets of the atom for the benefit of mankind. The political obsessions with power and regional hegemony that subverted their scientific ingenuity into the creation of the most frightening weapons of mass destruction the world has ever seen. The post-war development of civil nuclear power; its triumphs and tribulations, its rise and fall, and its resurgence in the opening years of this century as a possible energy option that may provide part of a carbon-free solution to climate change.

Parallel throughout is the relationship between Britain and Ireland: an Ireland that remained locked in a cycle of economic dependence on Britain for generations after it achieved political independence. For whom opposition to Sellafield's nuclear reprocessing industry and consequent contamination of the shared resource of the Irish Sea in time acquired an iconic status, becoming a credo of political probity across all parties, and providing the battleground for a sometimes real, sometimes phoney, war of words with Britain that culminated in a series of failed international court actions.

There are no heroes among these pages, only portrayal of all the normal traits of human nature: individual courage and idealism, naïveté and wisdom, cowardice and chicanery, political posturing, arrogance and stupidity, fear and anger, public duty, personal integrity and occasional examples of selfless commitment to the truth.

True to its purpose, this book can take no sides in the debate on the pros and cons of nuclear power. It simply lays out a narrative based on original research: from the archives of various media and public records, personal interviews and recollections of individuals, opinions and established facts. It is written in the hope of promoting a greater understanding of how we came to be where we are now, the warnings to watch out for on the political signposts we might follow, and the direction we may choose for our onward journey.

PART ONE

DISCOVERIES

1

New Beginnings

'Nothing in life is to be feared. It is only to be understood.'

Marie Curie

On 16 October 1927, Ernest Thomas Sinton Walton took the boat train from Dun Laoghaire to Holyhead in Wales and on to London. The twenty-four-year-old graduate of Trinity College Dublin (TCD) was heading for a new life as a physics researcher in the tranquil surroundings of Cambridge University in England.

The Ireland he was leaving behind was far from tranquil. The country had been partitioned five years earlier. Under the Government of Ireland Act, 1920, the Northern Ireland Parliament was inaugurated in June 1921, its six counties' border ethnically engineered to ensure a Protestant loyalist majority. The Northern unionist faction shrewdly wasted no time in setting up what originally, at least in the minds of the British government and the Irish nationalists with whom they were driven into negotiations in 1921 at the end of a guerrilla war, was envisaged as a temporary arrangement prior to full unification of a quasi-independent Ireland.

The unionists dug themselves into what rapidly became an entrenched and violent sectarian political establishment whilst the southern nationalists were temporarily distracted by a brutal civil war that raged from 28 June 1922 to 24 May 1923, between those who would settle for no less than a republic and those content with the stepping stones towards full independence offered by the terms of the Anglo-Irish Treaty that established the Irish Free State, Saorstat Eireann, in Ireland's remaining twenty-six counties.

The Civil War ended in the defeat of the republicans. Eamon De Valera, a veteran of Ireland's 1916 armed rebellion against the United Kingdom, leader and international lobbyist for an Irish Republic in the early years of the breakaway Sinn Féin administration of Ireland from 1918, now found himself demoted from statesman to nominal political head of the embittered remnants of Republican Sinn Féin and the IRA.

De Valera faced indefinite political isolation. The republicans continued to abjure the treaty and refused to participate in the new Irish parliament, the Dáil, which required an oath of allegiance to the British crown. In 1926, seeking to inject renewed vigour into a career and a republican cause that might otherwise dissipate into political irrelevance, De Valera founded Fianna Fáil, breaking with those elements within the old IRA, who refused to acknowledge the *de facto* existence of the Free State and the political necessity to become accommodated to it. A year later he led his followers into the Dáil.

In the latter years of one of the longest political careers of any statesman in twentieth century Europe, De Valera, increasingly obsessed[1] with justifying his place in history, and particularly some of the more controversial decisions of earlier times, insisted that neither he nor any of his republican deputies had formally taken the hated 'oath'. Instead, on physically entering the Dáil – the only way, De Valera claimed, for them to discover if in fact an oath was actually administered – the Clerk of the Dail had said to him: 'I only want your name in this book.' It was not for nothing that De Valera was described by political opponents as 'far too slippery to be caught by the naked hand'.[2]

With the Dáil seats of the republicans remaining vacant until late 1927, government of the fledgling Irish Free State was left to Cumann na nGaedheal, the Sinn Féin majority whose decision to accept the treaty had been publicly endorsed, at least in their terms, in successive general elections. Under the leadership of William T. Cosgrave, the Cumann na nGaedheal party operated the process of parliamentary democracy variously with the Labour Party, the Farmers' Party, the National League Party, a rag bag of independent deputies and the representatives of Dublin University who held reserved seats, providing a parliamentary opposition to the Cumann na nGaedheal executive.

As an agricultural country, Ireland, and all shades of its dominant farming community, had fared well during the Great War. Britain provided a relatively unlimited market for food, however varied or undistinguished its quality. Irish agricultural prices trebled between 1913 and 1920.[3] The collapse of the agricultural boom and any economic dislocation accompanying it were overshadowed by national political events surrounding the War of Independence, followed by the treaty and the Civil War.

The Cumann na nGaedheal government of the Free State tended to congratulate themselves on their perception of Ireland as an object of envy among other European nations.⁴ Defeating the republicans, however, had cost about £50m, the equivalent in today's terms of about nine billion euro, and left the newly independent state almost bankrupt.⁵ Beneath a façade of prosperity, the early Free State was strapped for cash and bound to a policy of fiscal austerity.

Emigration, long described as Ireland's economic and social safety valve, had resulted in the exodus of over two million Irish men and women between 1850 and 1910.⁶ Virtually ground to a halt during the years of the Great War, emigration now resumed its familiar pattern. During the first five years of the newly independent Irish nation-state an estimated 100,000 men and women departed Ireland for destinations in Britain or the United States. Parallel to this outward-bound tide of people who had been effectively reared for export,⁷ and whose departure warranted little hope or expectation of ever making a return journey, was a long-established pattern of seasonal migration to Britain. Thousands of Irishmen, women and children took menial jobs in agriculture, construction and industry in England and Scotland each year to supplement an otherwise meagre family income that barely rose above subsistence level.⁸

* * *

Walton was hardly typical of his fellow passengers on the Holyhead boat train. Comparatively speaking, his background was privileged. Irrespective of personal success in his chosen field, a comfortable middle-class future was reasonably assured, whether in Ireland or abroad. Walton was born in Dungarvan, Co. Waterford, in 1903, the first child of the Reverend John Walton, a minister in the Methodist Church, and Anna Sinton, whose farming family had been based for some 200 years in South Armagh, bordering the so-called 'linen triangle'.

Even at the turn of the century, Ireland's Methodist community was among the smallest of the country's Protestant churches, numbering about 17,000 in the southern counties of Ireland. The radical message of Methodism, with its principles of equality and tolerance, had challenged British religious orthodoxy in the eighteenth and nineteenth centuries. The founder of Methodism, John Wesley, visited Ireland on twenty-one occasions between 1747 and his death in 1791 and 'never found a field where people listened more eagerly to his preaching than in Ireland'. Methodist strongholds tended to develop along the main trade routes accessible to travelling preachers.

The Walton family traced its origins to one such stronghold, Clara, Co. Offaly, where John Walton, born in 1816, worked as a flour miller. The marriage of his son Samuel to Martha Woods in 1871 produced eight children. Their eldest son, Ernest's father John, a graduate of Trinity College, became a minister. The Methodist practice of short tenure for parish appointments, the 'three-year rule', meant that, by age three, Walton's family had moved from Dungarvan to Rathkeale, Co. Limerick.[9] In 1906, about a year following the birth of her second child, Dorothy, Anna died. Late in his own life, Ernest Walton recounted to his children a memory of being brought to his mother's bedside to see her shortly before she passed away.

Walton's formative years passed in a predominantly female environment amongst his mother's relatives in Armagh. As a small child, his favourite toys were tools with which Walton amused himself building and dismantling various gadgets. With his father's second marriage, the family was reunited and in 1914, at age eleven, Walton was sent to board at the Methodist College in Belfast.

Ireland's Methodist community, or 'swaddlers' as they were disparagingly called, lacked the political links with the establishment that characterised the Church of Ireland or the business and commercial connections of Presbyterians, and tended to remain aloof from political controversies. Nevertheless, the Methodist Conference in 1886 had voted overwhelmingly for maintenance of the union with Britain and rejection of the first Home Rule Bill. The Reverend John Walton, ministering in Cookstown, Co. Tyrone, in 1912, found himself dragged into the renewed national turmoil over Home Rule when he refused, on principle, to sign the 'Solemn League and Covenant', a pledge to defend unionism against the imposition of Home Rule, if necessary by arms. John Walton further refused to allow the manse to be used to collect the signatures of his congregation.[10]

At school in Belfast, Walton excelled in mathematics and science.[11] There he also made the acquaintance of Freda Wilson, another child of the manse, prompting his schoolmates to pen an irreverent ditty around an acronym of his name, STEW, that combined observations on his studious disposition with his amorous inclinations and predicted the couple's future marriage. Walton's plan to travel to Dublin by train to take the Trinity Entrance Scholarship examination was thwarted by the outbreak of the Civil War in Dublin. When the examinations were again laid on later in the year he won a scholarship. Combined with a County Scholarship from Armagh, the Trinity award provided the young Walton with sufficient income to sustain him throughout his college years.

Walton discovered that the physics training he had received at the Methodist College was superior to anything on offer in Trinity. By the early years of the twentieth century, Trinity had slipped from its previous status of eminence to a scientific backwater.[12] In his first year at Trinity, Walton claimed, without a trace of arrogance and with simple factual accuracy, to have learned nothing new. TCD's physics department, such as it then was, comprised a Professor, William E. Thrift, two assistants and an attendant, obliged to meet the educational requirements of medical and engineering students as well as catering for pure science students. Throughout his undergraduate career, Walton accumulated six academic prizes and was awarded first-class honours in his primary degree in mathematics and experimental science in 1926. He began research in hydrodynamics for an MSc in mathematics, but his heart lay in experimental physics, particularly recent discoveries concerning the atom.

* * *

The Great Exhibition of 1851, the showcase for Victoria's England, her empire, and her consort, Prince Albert, as well as rated a great public success, netted a substantial profit. Part of the profit was invested in a scholarship scheme, beginning in 1891, open to postgraduate students of British universities and throughout the empire. In 1895, another 24-year-old scientist, Ernest Rutherford from New Zealand's South Island, became an early beneficiary of the Imperial Scholarship's largesse, which brought him to the Cavendish Laboratory at Cambridge that October. Rutherford planned to continue his work on the detection of radio waves using a crude radio receiving device he had constructed in New Zealand. It would, he hoped, bring him fame and fortune. Rutherford's device preceded Marconi's perfection of radio transmission systems that would revolutionise the world of mass communications within a few years.[13]

Conducting experiments that involved passing electrical current through gases at low temperatures, the German physicist, Wilhelm Roentgen, noticed a fluorescent screen glowing at a distance from the glass cathode-ray tube he was using.[14] Roentgen called the unknown rays 'X-rays', since he had no other way of describing them, and did not know what was causing them. But he was in no doubt as to their power: when he placed his hand between the tube and the screen he was startled by a clear image of his own bones. Nor was he blind to their possible application: X-rays would prove of enduring value to medical science.

A month after his arrival at the Cavendish, Rutherford abandoned radio waves to begin research on X-rays with the Cavendish Director, J.J. Thomson. Two years later, in 1897, Thomson announced the discovery of a sub-atomic particle, the electron. On a practical level, work with cathode ray tubes would lead to the development of the modern television set. Scientifically, demonstrating that the electron existed showed that the concept of the atom, around since the time of Ancient Greece, was valid. It also disproved the belief that atoms were hard indivisible objects and the smallest units of matter.

Henri Becquerel, who held the Chair of Physics at the Museum of Natural History in Paris, heard of Roentgen's discovery at a meeting of the Academy of Sciences in Paris in January 1896. Intrigued, he began a series of experiments to see if Roentgen's results could be replicated using a variety of different substances.

Becquerel finally selected uranium salts for his experiment. It involved sprinkling the salts on a photographic plate that had been sealed in black paper and was then later developed. If the experiment was a success, the photograph would carry an image of the salts. Initially, Becquerel believed sunlight was needed to act as a catalyst. His first experiment produced the expected result, but when he was about to repeat it a few days later the Paris skies had clouded over. Becquerel shut the plate away in a drawer. On 1 March 1896, he developed the plate, expecting to find nothing. Instead he was surprised by an even clearer image. Becquerel had discovered what Marie Curie would later name radioactivity and what Rutherford was fated to spend the rest of his life and career exploring: probing and dissecting the atom to nudge out and reveal its remaining secrets.

At the end of the nineteenth century, there were some 1,000 physicists in the world, of whom about one hundred were engaged in atomic physics, experimental as well as theoretical.[15] In the early years of atomic research, no great distinction was made between physics and chemistry. The scientists knew one another well, and although personally and nationally they were fiercely competitive, results were shared freely at international conferences and through scientific journals. Like wandering journeymen, scientists migrated among laboratories and universities, working together on problems that attracted their mutual interest, often in a master–apprentice-type relationship.

* * *

The periodic table of pure elements was first published in 1869.[16] Basic chemistry, which allowed molecules to be broken down into their constituent elements and for those elements then to be weighed separately, had identified hydrogen, with an atomic weight of 1, as the lightest of all elements and uranium, atomic weight 238, as the heaviest.

The elements provided the raw material of the new physics. Within a year of publication of Becquerel's findings, Marie Curie proved that the electrical charge from Becquerel's rays was directly equivalent to the amount of uranium used in each sample. Extending her research to the next heaviest known element, thorium,[17] she demonstrated the same phenomenon. Experience of the dense uranium ore, pitchblende, led to a suspicion that it concealed an entirely new element, much more active than uranium.

Marie, with her husband Pierre, began the onerous task of breaking down the pitchblende to isolate the new element. Several months of back-breaking physical work boiling up pitchblende in a shed at their home in Paris were rewarded in the discovery first of polonium, named after Marie's native Poland, and, six months later, with radium.[18] In the paper announcing their discoveries in 1898, the Curies also noted that radioactivity seemed to be an inherent characteristic of the atom.

From his study of the radiations emitted by uranium and thorium, Rutherford identified two types, which he named Alpha and Beta, from the first two letters of the Greek alphabet.[19] The impecunious Rutherford was anxious to marry his New Zealand fiancée, Mary Newton. In 1899, just as his latest results on the nature of radioactivity were about to be published, he was offered the newly created post of Professor of Physics at the McGill University of Montreal. The post carried a salary of £500 per annum, enough for Rutherford to at last fulfil his long-standing engagement to Mary. The couple finally married in 1900.

In Montreal, Rutherford began a series of experiments on thorium and small samples of radium donated by Marie Curie. At an early stage he noticed that thorium gave off a gas whose radioactivity progressively reduced. The English chemist Frederick Soddy, who came to Montreal to work as a demonstrator in 1901, teamed up with Rutherford. In a series of papers over the next few years, Rutherford and Soddy explained the process of radioactive change that occurs in elements.

The scientists described how atoms could naturally decay, changing from one element into another; thorium to radon, for example, or uranium ultimately to lead. This rate of decay could be measured in 'half lives', the time it takes for half the atoms of one element to transmute into the atoms of another entirely different element or a physical variant or 'daughter' of the same element, which Soddy later termed isotopes.[20] Their 1903 paper 'Radioactive Change' provided the first informed calculation of the amount of energy released through this process of radioactive decay.

'The energy of radioactive change must therefore be at least twenty thousand times, and may be a million times, as great as the energy of any molecular [i.e. chemical] change,' they noted.

Addressing the Corps of Royal Engineers in the UK in 1904, Soddy remarked that all heavy matter probably contained enormous quantities of latent energy. The man who unlocked this store of energy 'would possess a weapon by which he could destroy the earth if he chose'.[21]

Einstein's special theory of relativity, published in 1905, had already supplied a mathematical formula to calculate this latent energy. No material can be lost from the universe. It follows that any atomic mass lost in the transmutation of one element to another is converted into another material form or released as energy. Einstein proposed that matter and energy are interchangeable; summed up in the equation $e=mc^2$; where e stands for energy, m for the mass of an object and c for the enormous figure that is the speed of light.

Neither Rutherford nor Soddy believed it possible to harness or artificially control atomic energy. A more immediate application of their discovery lay in measuring the rate of radioactive decay of elements and minerals that enabled calculation of the age of rocks and ultimately of the earth itself. Carbon dating, in which carbon 14 is used to estimate the age of materials of biological origin, would also prove particularly useful in determining the age of archaeological artefacts or, for example, sediments deposited during previous ice ages.

Rutherford's attention turned to the structure of the atom. Thompson's model supposed that the atom was like a pudding, randomly studded with electrons. From experiments at Montreal and later in Manchester,[22] where he returned to take up the Chair of Physics in 1908, Rutherford concluded there must be a central electrical charge in the atom around which electrons were possibly arranged in a spherical pattern.

Rutherford announced the discovery of the atomic nucleus to a stunned audience at a meeting of the Manchester Literary and Philosophical Society on 7 March 1911, attended mainly by local businessmen, a few young scientists and curious members of the general public. The first item on the society's agenda for the evening was a report from a fruit importer on the discovery of a rare snake in a bunch of Jamaican bananas.[23] Next came Rutherford's presentation, explaining that at the core of the atom there was a tiny but immensely powerfully charged nucleus. Much of the rest of the atom, its outer shells of electrons, Rutherford suggested, was largely made up of empty space.

If Rutherford's new model was correct, the atom was inherently unstable, because the negative electrons would be attracted by the positive nucleus and the structure would collapse. As the physical existence of the world and everything in it shows, atoms, in fact, are remarkably stable.

Niels Bohr, a young Danish theoretical physicist, arrived at Rutherford's Manchester laboratory in 1912. Rutherford set Bohr the task of resolving the contradiction that plagued his atomic model. Bohr's solution was ingenious. Bohr proposed that electrons occupied precise orbits, with precise energies, and that the number in each orbit was limited. If an electron falls from an outer orbit to a vacancy in a lower orbit, the difference in energy is emitted as a photon. Confirmation of Rutherford and Bohr's theory, and with it the new model for the atom, came in August 1913 from experiments in X-ray crystallography by their colleague in Manchester, Harry Moseley.

The world was enthralled by the spate of new scientific discoveries. The Hungarian chemist Michael Polanyi,[24] who later became a renowned philosopher of science, suggested that each great discovery acted as a 'growing point' to the next, akin to a process of cumulative intelligence. Using the analogy of a gigantic jigsaw puzzle, where each piece is held by a separate individual, Polanyi pointed out that the only way such a puzzle could ever be successfully completed is by each individual closely observing the work of the rest.

'Let them work on putting the puzzle together in the sight of the others,' he wrote, 'so that each time a piece of it is fitted by one, all the others will immediately watch out for the next step that becomes possible in consequence.'[25]

Physicists such as Roentgen, the Curies and latterly Einstein, some more reluctantly than others, became media celebrities. The discovery of radioactivity also fuelled entire new lines of business: for the average snake-oil salesman it was a must that every cure-all concoction contained radium, a guarantee of rapid sales irrespective of any curative properties. Radium, it was claimed, could turn grey hair back to black, without any dye. In Germany, toothpaste containing thorium was marketed as a surefire means of whitening yellowed teeth.

By far the most important application of the new physics came in medicine. Apart from X-rays, radium was used from the early years of the twentieth century in the treatment of cancer. In Ireland, radium was first used in or around 1904, when Dr John Joly, Professor of Geology and Mineralogy at Trinity College, presented four milligrams of a radium bromide to Dr Walter Steevens, the founder of Steevens Hospital in Dublin, to treat a patient's skin tumour.

Joly went on to pioneer the 'Dublin method' of radium therapy, using cross rays to concentrate the focus of radium treatment on a single point. In 1914, he suggested that the Royal Dublin Society establish a Radium Institute, mainly to co-ordinate supplies of the scarce and expensive element. Curiously, the Radium Institute was based in Leinster House, then the RDS headquarters. It was moved to Ballsbridge in 1922, when Leinster House was selected as the seat of the new Irish parliament.[26]

That over-exposure to radioactivity might also carry significant risks to health was lost on the media and the general public, though scientists were increasingly aware of the dangers. At a party in Paris in 1903, held to celebrate Marie Curie's doctoral award, Rutherford noticed the damage that experiments with radioactive materials had already visibly wrought to the hands of her husband, Pierre.[27]

Pierre Curie was killed in a traffic accident in Paris in 1906, his head crushed by a runaway horse-carriage. Harry Moseley died less than ten years later, the target of a Turkish bullet in the Dardanelles. The scientific community was scattered during the Great War, some dragged into the war effort, with others, like the pacifist Einstein, refusing to become involved. James Chadwick, a young Cambridge physicist who was on a working holiday in Germany when the war started, was stranded and held as a prisoner of war for its duration.

The war in Europe that cost the lives of nine million young men carried a portent of the weapons of mass destruction, the inducement of psychological terror on the battlefield and the deliberate bombing of civilian targets that would so distinguish the next major war of the twentieth century. On 25 May 1917, German planes conducted the first strategic bombing attack on a civilian target in an air-raid on Folkestone, Kent, killing ninety-five men, women and children and injuring another 195 people. A 2,000-lb bomb, the largest aerial bomb used in the Great War, was dropped on London on 16 February 1918.[28]

Modern science made its own unique contribution to this ghastly orgy of mass destruction in the formulation and military use of poison gas. While the gas[29] killed a relatively small number of fighting men –

about 30,000 – it disabled or horribly injured an estimated further million combatants, including an Austrian corporal, Adolf Hitler. The main effect of poison gas was psychological, especially the fear engendered among soldiers by the use of such indiscriminate and previously unknown weapons on the battlefield.

* * *

The pace of discovery in nuclear physics stalled during the war years. In between wartime duties, mainly on the development of radar to detect air raids, Rutherford continued experiments at Manchester and by 1917 succeeded in chipping particles off the nucleus. Within two years he proved that the atomic nucleus contained a positively charged particle, the proton, nearly two thousand times heavier than the lightweight, negatively charged electron in the atom's outer shells. Rutherford suspected that there was a further undisclosed particle hidden within the nucleus, a second particle without any charge, which he called the neutron. Experimentally, there was, as yet, no way to prove the existence of this neutral particle.

Rutherford had discovered the nucleus; he had described a model of the atom; in 1908 he had been awarded the Nobel Prize in Chemistry[30] for his work on radiation change, and a year later he received a knighthood. By the time he was nominated as Director of the Cavendish in 1919 as Thomson's successor, he was internationally acclaimed within the scientific community. He and the Cavendish held a magnetic attraction to aspiring young physicists from all over the world.

* * *

Working on his MSc in Trinity in 1926, Walton had learned that an 1851 Exhibition Overseas Scholarship might be available to one student from the Irish Free State in 1927. 'I was anxious to get to Cambridge which was *the* place to go to because there were so many famous physicists there at the time and of course Cambridge had a long history of mathematics and physics,' he said later.[31]

Walton applied for an 1851 scholarship, specifying an interest in atomic physics. However, his application came to Rutherford only very late in the day when the intake of postgraduates for 1927 was already complete. By letter, Rutherford advised Walton that there was no place available for him that year in the Cavendish, but didn't entirely close the door on his application. By Walton's own account, the difficulty was brought to the attention of Professor John Joly, who more than a decade earlier had collaborated with Rutherford on experiments to estimate the age of rocks.[32] Joly's intercession with Rutherford tipped the scales in Walton's favour. On 17 October 1927, Ernest Walton arrived at the Cavendish, in whose laboratories he would spend the next seven years of his life.

2

Splitting the Atom

'Split the atom's heart, and lo!
Within it thou wilt find a sun.'

From the writings of the Bahai' prophet, Baha'u'llah

Ebullient, loud, and brash, Rutherford, whose volcanic temperament matched his equally volcanic pipe-smoking habit,[1] might have been expected to intimidate the diffident, shy Walton, self-effacing almost to a fault. But from the beginning, the distinguished professor and the gentle-natured researcher showed a healthy respect for one another.

Meeting Rutherford to discuss suitable research projects, Walton suggested a new line of assault on the nucleus of the atom. Concentrated into a beam, electrically charged particles might have the potential to penetrate the strong electrical charge that physicists believed protected the nucleus. Instead of using slow-moving alpha particles derived from naturally occurring radioactivity, it might be possible to artificially generate a stream of the lighter, fast-moving electrons. Walton was unaware that Rutherford himself had only recently put forward much the same idea at a meeting of the Royal Society in London. Rutherford was enthusiastic about the proposed experiment, but first Walton was directed to the workshop where, among other things, he would learn how to blow his own glass for use in experimental apparatus. Moreover, he was warned to purchase his own supplies of raw materials.

After a slow start, America was fast catching up on Europe in the exploration of atomic and nuclear physics. In 1920, there were about 800 physicists in the United States; by 1932 they numbered 2,500.[2] Among them, J. Robert Oppenheimer later became the scientific head of the Manhattan Project to build the first atomic bomb. Ernest Orlando Lawrence, experimentalist and inventor of the cyclotron particle accelerator and later a pioneer of the hydrogen bomb, was only narrowly beaten by Rutherford's Cavendish team in the quest to split the atomic nucleus using artificial radiation.

The three great centres of experimental physics in Europe after the Great War were the Cavendish, the Curies' Radium Institute in Paris and Goettingen in Germany. In Copenhagen, Niels Bohr established the Institute for Theoretical Physics in 1921.

The 1920s was the heady decade of theoretical physics. Bohr united the laws of physics and chemistry in 1922, introducing new coherence to the periodic table: elements that are chemically similar are so, he said, because they have the same numbers of electrons in their outer shells. The Austrian physicist Erwin Schrödinger proved that electrons behaved as waves.[3] Later that year, Werner Heisenberg, a young Bavarian physicist, attended a lecture by Bohr at Goettingen. Building on Bohr's 'growing point', Heisenberg and his colleagues, Max Born and Pascual Jordan, developed a new system of mathematics, quantum mechanics, to describe the unpredictable behaviour of wave/particles at the atomic level.

At the fifth international Solvay Conference in October 1927, twenty-nine of the world's foremost physicists gathered to debate quantum mechanics. Eventually, after much agonising and debate, it was accepted that waves were particles and particles were waves, and that what Heisenberg termed the 'uncertainty principle' made it practically impossible to discern the position of electrons within the atom with accuracy at any given point in time. Albert Einstein, whose 1915 general theory of relativity showing how the universe worked on the larger scale had by now made him the most famous scientist in the world, remained sceptical of the 'uncertainty principle' to the end of his life. 'God does not throw dice,' Einstein famously declared. To which Bohr retorted: 'Nor is it our business to prescribe to God how he should run the world.'[4]

* * *

In experimental physics, of which the Cavendish was then the international centre, the cupboard of discovery looked increasingly bare. The Cavendish was famed for its reliance on bits of string and sealing wax to hold experimental apparatus together. The most prestigious centre of experimental research in the world was a monument to recycling – nothing was ever wasted, and anything that could be used again was carefully stored for the next experiment. Walton was surprised to find the Cavendish apparently as poverty stricken as Trinity. Years later, he charmed Irish audiences with lectures on his Cambridge days, peppered with stories of its legendary penny-pinching director.[5]

When some atoms are bombarded with protons, the radiation shows up as minute flashes of light on a fluorescent screen. Counting these scintillations was extremely laborious: eminent physicists sat for long hours in a box in a darkened space manually noting the frequency with which sparks appeared on screens in front of them. In 1908, while at Manchester with Rutherford, Hans Geiger had developed a relatively primitive machine for counting scintillations. But Rutherford would only countenance its use to confirm the accuracy of counts that had already been visually performed.[6]

Across the Atlantic, awareness of the military and technological importance of physics research ensured that American universities and laboratories were heavily endowed and supported by industry. The downside of corporate patronage, however, was that companies pressurised the American scientists for inventions and breakthroughs with immediate commercial application, to the detriment of pure research.

Walton's first major experiment, an attempt to construct a circular electron accelerator, was a failure, though it laid the groundwork for the development in America in 1940 of the betatron, used in the production of medical X-rays. His next area of exploration, to develop a linear acceleration tube for electrons, was also cut short when he came across reports in a German academic journal in early 1929 of similar experiments by Swedish and Norwegian scientists showing they were clearly ahead of him.[7]

Walton shared laboratory space at the Cavendish with John Cockcroft, an electrical engineer turned scientist, also from the north of England. The son of a small mill owner on the Lancashire–Yorkshire border, Cockcroft was some six years older than his companion. His training as an engineer had been interrupted by the Great War: he fought in France on the western front as an enlisted soldier with the Royal Field Artillery. In early 1918, to the great relief of his family, he became a commissioned officer and spent the remainder of the war in England. Cockcroft worked with the engineering firm, Metropolitan Vickers,[8] in Manchester before coming to Cambridge. At the Cavendish, when he was present in the laboratory, a rare enough event, Cockcroft was working on the interaction of molecules with metal surfaces.

'A very useful man,'[9] Rutherford once described Cockcroft, who had a genius for scrounging laboratory supplies from unusual places and negotiating deals with his industrial contacts to supply machines for research apparatus at knock-down prices. From his first days at

Cambridge, Cockcroft had become a very busy man and, like all busy people, his effectiveness in getting things done merely served to attract further work. He was assigned an increasing multitude of administrative and supervisory tasks by Rutherford.

Cockcroft also possessed the instincts of a theoretical physicist. He was attracted to a theory put forward by the Ukrainian scientist George Gamow, who for a time had visited Cambridge to work with Rutherford. Gamow had used quantum mechanics to show how particles behaving as waves would not only leak out of the nucleus of the atom, but might also be used to tunnel through its protective barrier to get inside it. Inspired by Gamow's wave theory, Cockcroft surmised that instead of accelerated electrons charged at millions of volts, an assault on the nucleus might succeed with the much heavier protons accelerated at a lower charge of 300,000 volts. In late 1928, Cockcroft discussed his calculations with Rutherford. Within days, Cockcroft and Walton were assigned a new project.

* * *

James Chadwick returned to the Cavendish following his wartime imprisonment in Germany and resumed experiments to capture the elusive neutron. By the late 1920s the scientific community had all but lost interest in Rutherford's neutron. Its existence was simply discounted. But the extraordinary results of an experiment conducted in Paris, by Irene Joliot Curie, daughter of Marie, and her husband Frederic, were about to breathe new life into Chadwick's quest.

The Joliot Curies were conducting tests on beryllium, bombarding the element with gamma rays. In January 1932, they reported their results in the French physics journal, results both Rutherford and Chadwick dismissed as impossible. The Joliot Curies concluded that irradiating beryllium with gamma rays resulted in the emission of high-velocity protons. Chadwick recognised that this was unlikely: gamma rays could deflect lightweight electrons, but protons are 1,836 times heavier than electrons. To his mind, only a particle of similar mass to a proton could have the effect observed by the Joliot Curies.

Over the next three weeks, Chadwick replicated the Joliot Curies' experiment, taking it several steps further. By 17 February, he had sent off a brief report to *Nature*, describing a new elementary particle, the neutron.[10] Carrying no electrical charge and with massive penetrative power, the neutron provided the key to finally unlock nuclear energy and create an atomic bomb.

* * *

From late 1929, Cockcroft and Walton were absorbed with the design and final construction of an apparatus to accelerate protons. Much of the design and building work was left to Walton, while Cockcroft sought out the hardware, including commissioning construction of a transformer from his old firm, Metro Vicks. Walton's creative dexterity and ingenuity were essential to the project.

'A good deal of the constructional work fell to me and I quite enjoyed this,' he recalled. 'I always enjoyed working with my hands.'

A lot of the experimental work fell to him too, since Cockcroft was often otherwise engaged. The first apparatus was complete in the spring of 1930, and Walton and Cockcroft managed to conduct some early inconclusive experiments before their transformer failed. By then, Walton had held the 1851 scholarship for three years and there was no possibility of any further renewal. Rutherford, resourceful as ever, recommended the Irishman 'as a man of exceptional ability' for a Department of Scientific and Industrial Research scholarship.[11]

At home in Northern Ireland visiting his family that summer, Walton took a train trip to Dublin. The train broke down and the passengers alighted at a small country station to wait for a replacement engine. Among the crowd on the platform was Freda Wilson, whom Walton had not met since their schooldays together in Belfast. Freda had trained as a Froebel teacher and was employed at a school in Waterford. They parted on a promise to write.

Walton could build machines and, as one of his Cavendish colleagues noted, probably fashion the smallest parts of a watch, but love letters were clearly beyond him. For the first four months of an unremittingly dull correspondence, he addressed her as 'Dear Miss Wilson', whilst she wrote back, tongue in cheek, to 'Dear Stew', only dropping the hated nickname in favour of Ernest when he agreed to address her in future as Freda.[12] In the autumn, Freda briefly visited Walton at Cambridge and they arranged to meet again in Belfast over the Christmas holidays.

At the Cavendish, Cockcroft and Walton, with Rutherford's support, had decided to build a new machine capable of generating up to at least 700,000 volts. They were influenced by reports from America of Lawrence's cyclotron and other particle accelerators producing beams of one million volts. Cockcroft was plagued by doubts about his own calculations based on Gamow's wave theory. The experimenters were no longer sure that 300,000 volts would be enough.

Walton was also busy with his doctoral thesis. In mid-1931 Cockcroft took off on a four-month tour of the Soviet Union. When he returned in September, Walton was in Ireland but construction of the new apparatus was well advanced. By January 1932, the machine was ready, though their first experiments yielded little. Early in April, under pressure from Rutherford and Chadwick, they began experiments using the element lithium.

Beneath the array of gigantic glass bulbs vacuum-sealed with plasticine, and tubes and transformers that reached as high as twenty feet, Cockcroft and Walton had installed a small observation hut, lined with lead to protect them from radiation. To reach the hut meant crawling along the floor when the apparatus was switched on, to avoid being electrocuted. On 14 April, Walton was working alone in the laboratory. He cranked up the machine to produce a stream of protons accelerated to 700,000 volts, then crawled on his hands and knees to the observation hut to see if anything was happening.

'When I looked in through the microscope I could see a whole lot of little stars suddenly appearing and disappearing,' he remembered.[13]

He tried switching on and off the apparatus a few times, then realised he was not imagining the effect. Walton called Cockcroft, who confirmed his observations.

'We managed to get Rutherford into this little hut. He was a big man and it took some manoeuvring to get him in to sit on the low stool,' Walton explained. 'Then he issued various instructions like "increase the accelerating voltage" or "increase the proton current" and so on. Finally he told us to shut down the apparatus.'[14]

Rutherford was in no doubt what he was looking at: he was entirely familiar with alpha particles since he had named them in the first place.

A proton has a mass of one on the atomic scale. Lithium has an atomic mass of seven. Injected with a proton, the lithium atom split apart, disintegrating into two alpha particles, or helium nuclei, each with an atomic weight of four and each with energy of eight million electrons. Curiously, the effect was apparent even at a voltage of 125,000, which meant the larger apparatus they had constructed over the previous eighteen months was unnecessary.[15] Cockcroft and Walton had split the nucleus; they had confirmed Gamow's quantum mechanics theory and, for the first time, verified Einstein's formula, $e=mc^2$.

Over the next ten days, as well as preparing an announcement for *Nature*, Cockcroft and Walton tested their proton beam on other elements, including boron and fluorine, producing results similar to lithium.

Walton wrote to Freda of his 'red-letter day', 14 April 1932, and she replied with warmth and pride in his achievement. Rutherford chose a meeting of the Royal Society in London on 28 April to announce the splitting of the atomic nucleus. By May Day, the newspapers were at full throttle, and at the laboratory Cockcroft and Walton found themselves temporarily besieged by journalists and reporters.

The *Daily Mail* speculated that the discovery might lead to the transmutation of 'lead into gold'. In Ireland, national newspaper coverage was more muted, although Freda wrote to Walton with delight that the Waterford papers had claimed him as one of their own. 'I have learned one thing from reading the papers this week,' Walton wrote Freda, 'and that is not to believe all you see in them.'[16]

* * *

No doubt Eamon De Valera, newly elected President of the Free State Executive Council, would have empathised. At the general election in February 1932, De Valera's Fianna Fáil was returned as the largest party. With Labour's support, De Valera was successfully nominated as President when the Dáil reconvened the following month. About a year before the election, De Valera had launched a new national newspaper, the *Irish Press*, which was just as well since from his perspective the established papers, the *Irish Times* and *Irish Independent*, hardly ever had a good word to say about him. British media coverage reflected the widespread paranoia about De Valera embedded within the British establishment.[17]

'Fanatical, ruthless, self-centred and self-sufficient, this man is dangerous,' an editorial in Britain's Tory press described the Free State's new leader.[18]

Lord Granard, a member of the 1922 Irish Senate, declared: 'I have never met anybody like the President of the Executive Council of the Irish Free State before. I hope that the Almighty does not create any more of the same pattern and that he will remain content with this one example.'[19]

At the Eucharistic Congress celebrations in May 1932 – the 1930s equivalent of a papal visit – William T. Cosgrave, who had shepherded the Free State through its first ten years, refused to attend the official reception at the same time as De Valera.[20] A decade of independence had done little to dissipate the festering bitterness of the Civil War. Those who, until the break over the treaty, had been his comrades in arms in the struggle for independence reviled and distrusted their former chief. De Valera, they felt, had betrayed them. De Valera's Fianna Fáil was denigrated as 'slightly constitutional', as not having quite severed its umbilical cord to the IRA, whose elements were still bent on causing mayhem and civil unrest, in Britain as well as Ireland, and who were further tainted by associations with Russian communism. As for the man himself, if the red smear wouldn't stick then at the very least he was a threat to democracy and Ireland's international position as a constitutional state.

Successive British administrations in the late nineteenth century, partly in the hope of 'killing Home Rule with kindness', had imposed land reforms that distorted any later prospect for agricultural development by creating a whole class of small landowners on scarcely viable holdings. The British, furthermore, had left most of the education system, as well as the hospitals and healthcare, under denominational control, principally by the Catholic Church. The largely Protestant Irish ascendancy[21] was politically abandoned long before the nationalist takeover, and many ordinary Irish Protestants similarly felt like aliens in their own country. British policy towards Ireland ensured that the wheel had come full circle and the hegemony of a Protestant ascendancy was replaced by the dominance of Catholic nationalism.

One of the most curious and least studied phenomena of its first decade is the drastic reduction in the Protestant population of the new Irish state. The only mass displacement of an ethnic group within the British Isles since the seventeenth century, the Free State lost some 50,000, or 34 per cent, of its non-Catholic population in its first ten years.[22]

The Protestant exodus was influenced by a variety of things, including antipathy towards what many Protestants perceived as an emerging confessional nationalist state, which motivated many to either move to Northern Ireland or out of the country entirely. For those who lacked the means to go, there is evidence of suffering periodic religious discrimination, socially and in employment, though nothing that compared with the rampant sectarianism that characterised Northern Ireland. At grassroots political level and in the nationalist media, Protestantism was frequently, if wrongly, equated with residual loyalties towards unionism. The nefarious impact of the Vatican's 1904 *ne temere* decree, which obliged mixed-marriage couples to raise their children as Catholics, ensured that within a generation most of the Free State's Protestant population had been married out of existence.[23]

In their decade of power, Cumann na nGaedheal governed the Irish Free State conservatively. They inherited a well-oiled public service administration from the British, which adapted with ease to its new political masters. Cumann na nGaedheal provided political stability and, by contemporary international standards, the newly independent state weathered the times relatively well. Because of the nature of its agricultural economy, which relied mainly on cattle exports to Britain, the Free State was at least initially inured against the worldwide impact of the Great Depression of 1929 and actually benefited from lower prices for imported grain. America's boom years in the 1920s had attracted anywhere between 20,000 and 30,000 emigrants from Ireland annually.[24] But the effect of the worldwide depression eventually caught up, closing the emigration valve by 1930 and precipitating a balance of payments crisis for the government.

The numbers of unemployed in the Free State rose to 80,000 and beyond. The continuing lawlessness of the IRA, assassinations of key political figures and the uncovering, by the head of the civil police force, Garda Commissioner General Eoin O'Duffy, of a bizarre, so-called republican communist plot spurred the enactment of a highly unpopular Public Order Act, complete with military tribunals, to suppress so-called communist IRA elements.[25] With the economy foundering, the Cumann na nGaedheal government introduced tax increases and pension payment reductions in 1931, doubtless contributing to its loss of popular support in the following year's election.

De Valera's platform for the 1932 election centred on old republican idealism and repudiation of the Anglo-Irish Treaty. A Fianna Fáil government would excise the despised oath and other vestiges of loyalty to the British crown from the Free State constitution. De Valera's promise to withhold payment of land annuities due to Britain arising from the tenant purchase Land Acts of the late nineteenth and early twentieth centuries appealed to the farming classes.

Writing to Walton at Cambridge in mid-1932, the Provost of Trinity College, E.J. Gwynn, referred to 'the political disturbance in the Free State' and noted that 'the future position of this country is so uncertain'.[26] Such fears were misplaced: beneath his populist agenda, De Valera was just as firmly wedded to a conservative political order in Ireland as Cumann na nGhaedheal. As a politician, he was primarily committed to appeasing his own political constituency and assuaging their grievances. In establishing Fianna Fáil, however, he had made the break with those republican elements who opposed the adoption of a the British-style model of parliamentary democracy in Ireland and there was little to distinguish the broader economic and social aims of Fianna Fáil from those of the other main political grouping, Cumann na nGaedheal, and their shared commitment to protecting the stability of the Irish state's nascent democracy.

3

Darkening Clouds

'Anyone who expects a source of power from the transformation of these atoms is talking moonshine.'

Ernest Rutherford

It was to this transforming Ireland that Ernest Walton returned as a Fellow of Trinity College and lecturer in experimental science in 1934. Shortly afterwards, he was married to Freda Wilson in the Centenary Methodist Church on Dublin's St Stephen's Green in a ceremony jointly conducted by their fathers. Rutherford sent the happy couple a copper tray as a wedding gift.

As he settled into married life and his job as a lecturer in Trinity, Walton must have privately reflected on how Ireland had changed in the seven years he had been at Cambridge, as the new nation-state set about forging an identity separate and entirely distinct from its colonial past and chafed at the bonds of economic dependence, culture and the mixed ethnicity and shared language that tied it to Britain.

In the early years of independence, Anglo-Irish relations were reasonably cordial. The first Irish government's relations with Britain had been directed towards portraying Ireland as a responsible dominion within the imperial fold.[1] Independence may have been forged out of rebellion against the crown, but the nationalist Irish state's policy towards its former colonial master was intended to project respectability, a stability of which it felt less assured, and acceptance within the broader family of sovereign nation-states. To this end, by 1923, the Free State had joined the League of Nations.

Cosgrave's Cumann na nGaedheal government adhered to the letter of the terms of the Anglo-Irish Treaty. Britain, after all, was Ireland's largest trading partner and self-interest demanded acquiescence even to those parts of the treaty that stuck in the nationalists' craw. In the mid-1920s, agreement was reached with Britain on a financial settlement, the essence of which was that the Free State would no longer carry a burden in respect of Britain's national debt, but would continue to pay land annuities – the interest charges arising from the old tenant purchase schemes. While political independence was one thing, economic and fiscal independence proved quite another: apart from the overwhelming dependence on Britain for trade, the Currency Act, 1927, tied valuation of the Irish pound to sterling in an arrangement that persisted until 1978.

Personal rather than fundamental ideological differences characterised relations between Cumann na nGaedheal and Fianna Fáil. The nationalist parties' common origin in Sinn Féin meant they shared a common vision of an ideal Gaelic Irish state. In time, that Gaelic identity would become increasingly synonymous with the Catholic faith.[2] Within a set of static cultural presumptions, which probably owed more to nineteenth-century romanticism and the Celtic revivalism of Yeats and others in the early twentieth century than to cultural realities of the lives of most Irish people, restoration of the Irish language as the spoken vernacular had been adopted as a key national priority.

The education system was the silver bullet for the language revival. One of the early acts of the Cumann na nGaedheal government was to eliminate natural science subjects from the primary school curriculum in order to facilitate more time for the teaching of Irish.[3] Since by far the majority of Irish people finished their formal education at primary level, a whole generation was robbed of any educational foundation, even at the most rudimentary level of knowledge, in the sciences. The attempt to restore Irish as the spoken language, while itself a noble and laudable objective of 'de-anglicisation',[4] ultimately proved a dismal failure. The unacknowledged, and, for decades, unacceptable truth was that, throughout much of the country, the use of Irish in the vernacular had been replaced by English, sometimes as long as two to three hundred years before.[5]

Walton discovered that Trinity was just as impoverished as, if not more than, when he had left for Cambridge in 1927. A Royal Commission had proposed a grant of £113,000 and an annual subsidy of £49,000 for Trinity in 1920, but the funding plan lapsed with Irish independence. The new Free State government was hardly in any position to provide adequate financial support to its oldest university. Much later, Trinity would turn to De Valera for rescue.[6] As the world depression began to bite in Ireland, Trinity, like the rest of the state's institutions, was obliged to tighten its belt.

* * *

Apart from removing O'Duffy as head of the Gardaí,[7] De Valera's accession to the helm of the Irish Free State in 1932 did not herald the purges of the civil service or the army so fearfully anticipated by his opponents. Like Cosgrave before him, De Valera rapidly attuned himself to the demands of executive office, and the key figures of the civil service, in turn, swivelled their own hats backwards to amiably greet his arrival.

Relationships with Britain deteriorated quickly, however. As well as being President of the Executive Council, De Valera initially retained the portfolio of Foreign Affairs. An early decision to fulfil the manifesto promise to halt the payment of land annuities, totalling about £5m per annum, drew swift retaliation from the British side, who slapped tariffs on Irish beef and other agricultural imports as well as imposing quotas on imports from Ireland.

Negotiations failed to resolve the dispute. The British were agreeable to international arbitration, but De Valera refused to accept British conditions on the appointment of judges to a Commonwealth tribunal. Such a body, De Valera said he had told the British, would have 'a natural bias against Ireland', and any of its members who belonged to the States of the British Commonwealth would find it difficult 'to appreciate the historical and other background of the whole of the present position so far as Ireland was concerned'.[8] Ireland was plunged into the so-called economic war with Britain.

The economic war, the opposition castigated Fianna Fáil, had been presented as a struggle whereby 'the Irish people would grow so fat that we would have to put new doors in Irish houses in order to let them pass in and out'.[9] Instead, Irish exports declined by over 50 per cent between 1932 and 1934 and agricultural prices plummeted by a third of their value. Unemployment rose to 100,000, although the government was not inclined to accept the blame. 'Emigration has stopped from causes outside our control and the unemployment problem has been aggravated on that account,' was the preferred explanation of the Minister for Industry and Commerce, Sean Lemass.[10]

To compensate for the disastrous impact of the economic war, the Fianna Fáil government introduced a self-sufficiency policy, erecting strong tariff walls against imports to protect and encourage Irish manufacturing. Over time, ironically, this policy resulted in more import dependency on Britain than actual self-sufficiency in Ireland. A cattle–coal pact with Britain, signed in 1935, increased the beef export quota, but left the retaliatory tariffs' framework in place.[11] The pact was renewed in the following two years, presaging the

Anglo-Irish Trade Agreement of 1938 which brought the economic war to an end.

De Valera was the first leader of an independent Irish state to tap into nationalist popular sentiment and a residual strand of antipathy towards its former colonial masters that existed among the masses as a means both of promoting his own favoured policies and of masking their deleterious effects. The land annuities dispute with Britain was an economic disaster for the country, but politically its portrayal as a manifestation of Irish independence had its uses.

Likewise, De Valera could claim a popular mandate for systematically unpicking the political aspects of the 1922 treaty with Britain. Starting with a Dáil vote to remove the reviled oath of allegiance in 1933,[12] he introduced a raft of legislation to break all remaining crown connections. By 1937, De Valera had succeeded in passing a new constitution, which more accurately reflected the original nationalist prescription for an independent Ireland. While it stopped short of declaring a republic, the new constitution effectively removed any lingering vestiges of loyalty to the British crown. Its claim to the territory of the whole island of Ireland, however, enraged the Northern unionists.

If the Cumann na nGaedheal leadership greeted the unilateral dismantling of the treaty's provisions, which they regarded as an inviolable international agreement, with considerable anguish,[13] the British establishment was outraged. British ire was mainly directed towards the person of De Valera himself. Not for the first time, nor for the last, the British failed to understand the domestic political imperatives that guided De Valera's actions, or the popular appeal within Ireland of asserting full independence from Britain, which after all had been the signature tune of militant Irish nationalism since the late nineteenth century.

British policy towards the Irish Free State, and De Valera as its leader, soon mellowed. Chamberlain was keen to end the economic war and reach some settlement on Ireland's position within the empire and the Commonwealth. Top of the British agenda was a desire to keep the Irish state inside the imperial fold. In the event of war, the British held to a deluded belief that Ireland would rally to Britain's side. Top of De Valera's agenda was reunification of his country. Falling short of that, De Valera was determined that the Irish state's complete independence from the empire would be made apparent.

Had the British studied De Valera's public and parliamentary statements from 1936 onwards with any diligence, they could hardly have mistaken his intention to declare Irish neutrality in the event of war. As part of the 1938 settlement, De Valera secured the unconditional return of the treaty ports – in which Britain had maintained a strategic defensive presence under the terms of the 1922 Treaty[14] – to Irish control. More than any other event, the return of the treaty ports to the Free State copperfastened De Valera's ability to declare Irish neutrality in any future conflict.

Winston Churchill, in political limbo within his own party, reacted furiously when he learned the terms of the settlement with Ireland. Excoriating his government's action in the House of Commons, Churchill made his opinion, and mistrust, of De Valera abundantly clear:

> Now we are to give them [the treaty ports] up, unconditionally, to an Irish Government led by men – I do not want to use hard words – whose rise to power has been proportionate to the animosity with which they have acted against this country.[15]

Churchill had no illusions about De Valera's intention to keep Ireland out of any forthcoming war and what it could mean for the British war effort. A declaration of Irish neutrality at the start of the war meant use of the ports would be irretrievably lost to Britain, undermining its North Atlantic naval defences. The price of Irish neutrality, Churchill suggested, would be counted in British lives.

Churchill's warnings fell on deaf ears. Even at the later stages of the 1930s, there was little political or popular appetite in Britain, or throughout most of Europe, for any repeat of the cataclysmic conflict of the Great War. Appeasement, a policy of containment by negotiation, was preferred to direct confrontation with Hitler's predatory militarism. A newsreel from May 1938 showed the English soccer team in Berlin, albeit sheepishly and uncomfortably, giving the Nazi salute at the start of a match with Germany. Appeasement diplomacy reached down to the level of international sporting occasions.[16] De Valera strongly supported and encouraged Chamberlain's policy of appeasement in Europe. Any war involving Britain would only raise difficulties and introduce unwanted complexities into already fragile Anglo-Irish relations.

* * *

The rise of fascism in Germany had already scattered its scientific community to the four winds. Departing their German home in

December 1932, on a journey that would see him eventually settled at Princeton, Einstein told his wife, Elsa: 'Take a good look at it. You will never see it again.'[17]

Less than two months later, on 30 January 1933, Adolf Hitler was appointed German Chancellor. On 1 April, the Law for the Restoration of the Professional Civil Service was promulgated and lists published of Jews dismissed from their university posts. Rutherford, Einstein and Bohr rallied to the cause of the displaced scientists – many of whom were shocked to discover that crude ethnicity could so peremptorily extinguish their careers – fundraising and finding placements and jobs for them wherever possible in British, French and Scandinavian universities and laboratories.

Several of the cohort of young Hungarian physicists, including Leo Szilard, Eugene Wigner and Edward Teller, who had already fled the brutal Hortha fascist regime in Budapest, found themselves on the run a second time, this time from Germany. Over the next eight years, some one hundred physicists and their families fleeing Nazi persecution sought refuge in the United States.[18]

Next it was Italy's turn. In July 1938, Mussolini published the Manifesto del Razza, which solemnly pronounced: 'Jews do not belong to the Italian race.'[19] Italy's most well-known physicist, Enrico Fermi, immediately set in train plans to emigrate to America, with his wife Laura, who was Jewish, and their two young sons. Fermi seized on the opportunity of his award of the 1938 Nobel Prize in Physics to slip his family out of the country. Later that year in Helsingor, near Copenhagen, Niels Bohr addressed an international scientific congress on the horrors and incipient inhumanity of totalitarianism, making a plea for tolerance of diversity of race and culture. The German delegation walked out.

* * *

Despite the savage dress-rehearsal of 'lightning war' tactics played out in the Spanish Civil War and the consolidation in power of right-wing totalitarian regimes in Hungary, Italy and Germany, the political and military catastrophe that would destroy the lives of tens of millions of people across Europe was as yet experienced in ripples rather than waves. In Europe's laboratories and institutes, there was no indication as to the use that would ultimately be found for the discoveries in atomic science.

In a series of experiments in Paris in 1932, the Joliot Curies had shown how to irradiate matter or transform one element into another by bombarding it with particles, in the first successful demonstration of artificial radioactivity. They were awarded the 1935 Nobel Prize in Chemistry for this achievement, through which 'at last the old dream of the alchemists has become reality'.[20] Where the alchemists of old had been mainly motivated by greed, the Joliot Curies were honoured for the contribution which artificial radioactivity, and with it the ability to manufacture an array of different isotopes, would make to medicine, especially in the more effective treatment of various forms of cancer.[21]

The French discovery inspired Enrico Fermi and his team in Rome to start their own experiments bombarding elements with neutrons. Over the next two years, they proved that the presence of hydrogen behaved as a moderator, slowing down the speed at which liberated neutrons travel, which facilitates the penetration of atomic nuclei. It was all down to the use in Italy of marble tabletops – experiments carried out on wooden surfaces, with their hydrogen content, delivered different results to those where marble was the surface. Elements such as boron absorb neutrons and so reduce the number of nuclear disintegrations. The Fermi team's experiments showed that nuclear reactions might be controlled and manipulated.

In a 1937 paper, Niels Bohr proposed that the nucleus of an atom was like a drop of liquid, held together by what he identified as the strong nuclear force, opposed by the electrical charges of the positively charged protons. In natural elements, the forces within the nucleus cancel each other out. Bombarding the nucleus with neutrons caused it to wobble, Bohr suggested, and destabilised it, much in the same way as adding water to a glass generates motion in the liquid.

Inspired by Bohr's insight, teams in Italy, France and Germany began working separately on slow neutron bombardment experiments. By 1938, Otto Hahn and Fritz Strassman had identified a series of specific isotopes of uranium arising from their slow-neutron experiments. In search of a theoretical framework for his laboratory experiments, Hahn turned to his old colleague, Lise Meitner, with whom he had worked for many years.

Meitner, a Jew, had long since fled Germany with Bohr's help, for safe exile in Sweden. Over the Christmas holiday that year, Meitner and her nephew, Otto Frisch, endlessly discussed Hahn's results. Early in January, Meitner became convinced that Hahn and Strassman had burst the nucleus of the atom, what Otto Frisch later termed nuclear fission. Using Einstein's $e=mc^2$ formula, Meitner calculated precisely the amount of energy released in such a process at 200MeV.

* * *

Perverted genius, charlatanism and ego-mania all combined, more or less in equal measure, in the character of the diminutive, eccentric Hungarian physicist, Leo Szilard, who left Budapest for Berlin in 1919. Szilard was prone to flashes of insight, following long walks or long baths, as opposed to getting his hands dirty through experimental work. Once, having imposed on the reluctant hospitality of a friend and his wife, Szilard immediately demanded to know where the nearest hotel was, because the bed they offered was too hard to sleep in. When he stayed in hotels, cleaning staff complained of the state he left bathrooms in. When he thought about nuclear energy, Szilard was more inclined to consider its military rather than medical potential.

Szilard was irritated that Rutherford, in a 1933 interview, had dismissed the idea of extracting vast quantities of energy from the atom as 'mere moonshine'. Waiting for a London traffic light to change as he returned home from a walk around Bloomsbury, Szilard conceived the idea of a neutron chain reaction. Later he claimed inspiration came from the H.G. Wells novel *The World Set Free*, as much as from developments in nuclear physics.[22]

All that was necessary, Szilard reckoned, was to find an element that would emit two neutrons as it absorbed one neutron. Because of their neutral charge, the liberated neutrons would travel on and on, propagating a chain reaction and culminating in a massive spontaneous release of energy. A nuclear bomb was possible. Szilard promptly patented his idea.

Based on their knowledge of the atom and all the experimental data accumulated over the best part of half a century, neither Rutherford, nor Einstein nor Bohr believed in the mid-1930s, as Szilard had convinced himself, that it was practically feasible to build a nuclear bomb.

Rutherford would not live to change his mind or to have it changed for him. Following a routine hernia operation in October 1937, he developed septicaemia and heart failure and died within a few days. Writing to Robert Oppenheimer that December, Niels Bohr spoke of his old friend and mentor: 'Life is poorer without him, but still every thought about him will be a lasting encouragement.'[23]

PART TWO

THE BOMB MAKERS

4

Weapon of Mass Destruction

'A small amount of matter, the product of a chain of huge specially constructed industrial plants, was made to release the energy of the universe locked up within the atom from the beginning of time. A fabulous achievement had been reached.'

War Department Release on New Mexico Test, July 1945

Leo Szilard, who arrived in the United States in January 1938, had become convinced of two inevitabilities: war with Germany and a race among the belligerents to produce an atomic bomb. Logically, to his mind, it followed that all future experiments on nuclear fission should be kept secret.

From America, Szilard telegraphed Frederic Joliot in Paris in February 1939 pleading for secrecy on the results of any further neutron experiments. Although Szilard was unaware of this, Joliot, with his colleagues Hans Von Halban and Lew Kowarski, was due to publish results of experiments they had conducted that showed multiple neutron emissions.

Szilard's plea was firmly rebuffed. The Frenchman was not ready to abandon the tradition of sharing ideas and results that had for so long been intrinsic to the development of physics. Nor was he about to forgo the academic glory of announcing new discoveries. At the back of his mind, Joliot dreamt of building a nuclear reactor that would use a controlled chain reaction to heat uranium fuel which could then be harnessed to power a turbine and produce electricity. Joliot had already patented a basic reactor design.

Niels Bohr, too, abhorred the notion of scientific discoveries being shrouded in secrecy. Much of his life had been devoted to building up a community of scientists in which scientific theory and the results of experiments, good and bad, were openly published and freely shared. Similarly, Enrico Fermi initially resisted Szilard's apostolic mission of nuclear secrecy.

What the scientists failed to recognise was that the freedom to work together, learn from one another and openly share discoveries and insights, in which the international physics community had luxuriated for the best part of half a century, was only sustainable in a world where political and military administrations were indifferent to discoveries about how the world or the universe worked and could see no direct application for the new physics to the pursuit of war or power. Before long, physicists, irrespective of where they were located, would lose the freedom to make their own choices on what to publish or exert any real influence on how their discoveries might be applied.[1] Politics was at last about to catch up with them.

* * *

In April 1939, a few weeks after Hitler had swallowed up what remained of Czechoslovakia and as Mussolini, the conqueror of Abyssinia, prepared for his next military adventure in Albania, Joliot's French team published a second paper in *Nature*. Uranium bombarded with neutrons would fission, Joliot claimed. Provided a suitable moderator was in place, 'the fission chain will perpetuate itself,' the paper stated. Heavy water was identified as a moderator – so called because, although chemically identical to ordinary water, in heavy water the hydrogen atom is replaced by its heavier isotope, deuterium.[2]

Heavy water is difficult, dangerous and expensive to make. In 1938 the world's entire supply of about 180 kilograms was at the Norsk Hydro plant in Norway, where it arose as a by-product of ammonia manufacture. With his dream of an energy reactor in mind as well as a fear of what the Germans might do with a supply of heavy water, Joliot suggested the French government should purchase the entire stock from Norsk before, as he anticipated, the Germans could get their hands on it. When Norsk was approached about the deal in February 1940, the French discovered that the German company I.G. Farben, which owned 25 per cent of Norsk, had already demanded the heavy water be sent to Germany.[3]

The Norwegians offered to loan the heavy water to the French for free. In an escapade worthy of a James Bond movie, Lieutenant Jacques Allier of the French Secret Service fooled German agents at Oslo into thinking the consignment of twenty-six cans had been loaded on a plane to Amsterdam. In fact, he had slipped the heavy water and himself onto a plane bound for Perth in Scotland. Three weeks after the consignment finally arrived safely at Joliot's Paris laboratory, the Germans invaded Norway on 9 April 1940.

Within two months, Germany's *blitzkrieg* reached Paris. Fleeing south to Bordeaux, Joliot, van Halban and Kowarski, loaded the heavy water onto the British merchant ship the *Broompark*. At the last moment, Joliot decided to remain in France. His wife, Irene, had taken ill on the journey from Paris and he was reluctant to leave her behind. Van Halban and Kowarski sailed for Falmouth aboard the collier with its strange cargo. The Norwegian heavy water eventually found its way to Canada, and the Chalk River site in Ottawa, and was used in an experimental reactor built there to manufacture plutonium in 1944.

In the September 1939 edition of the US journal *Physical Review*, an article by Niels Bohr and the American physicist, John Wheeler, one of the last of its kind published throughout the remainder of the war years, identified uranium 235 (U235), an isotope present in small quantities in natural uranium 238 (U238), as a highly fissionable material. The obvious technical difficulty lay in extracting it from uranium 238. Most scientists, Bohr among them, believed it was practically impossible to extract U235 in sufficient quantity from U238 to generate a sustained fission chain reaction and ultimately create a bomb.

Like Szilard, and sometimes in an uneasy collaboration with him in Berkeley, Fermi was replicating the French experiments. By mid-1939, Szilard had worked out that carbon-based graphite might act as a suitable moderator for fission reactions in U235, in much the same way heavy water was known to do, and suggested as much to Fermi. By now he feared that neither Fermi nor anyone else was taking him seriously on any subject.[4] Back in New York in mid-1939, Szilard turned to his compatriots, Eugene Wigner and Edward Teller, for moral and practical support in his personal crusade to stay the assumed relentless march of Germany towards development of atomic weapons.

* * *

In Britain and America, Szilard's early efforts to press the case for atomic weapons development met with a response from the military establishments along Rutherford's lines of 'mere moonshine'.

Szilard was concerned that the Belgian government might be inveigled into selling uranium from its Congolese mines to Germany. This line of reasoning led to talk of Albert Einstein, who was known to be acquainted with the Belgian royal family. If Einstein could be persuaded to intervene, his status would add considerable weight to their political lobbying.[5] Over several days in New York, the Hungarian scientists also debated contacting the US administration in Washington. Several meetings later, in August 1939, a letter mainly drafted by Szilard, but signed by Albert Einstein, was provided to an intermediary, the New York economist Alexander Sachs, for personal delivery to the President, Franklin D. Roosevelt.

Einstein's letter urged the President to direct his administration to co-operate with the efforts US-based scientists were making in atomic research. Einstein warned, ominously, that: 'Germany has actually stopped the sale of uranium from the Czechoslovakian mines which she has taken over.' 'Some of the American work on uranium is now being repeated,' he wrote.

Events conspired to delay Sachs from making his presentation to the President. On 23 August 1939, Germany and the Soviet Union signed a non-aggression pact, which effectively carved up Poland between them, sealing its fate. Germany's *blitzkrieg* against Poland began on 1 September. Two days later Britain and France declared war on Germany. Sachs finally got to see Roosevelt in the middle of September and secured his personal commitment to follow up Einstein's letter. Although a month later the President replied by letter to Einstein that he had established a committee 'to thoroughly investigate the possibilities of your suggestion regarding the element of uranium', nothing much would come of it for a further eighteen months.[6]

* * *

Every large-scale human crisis, including war, provokes decisions that might otherwise be delayed or develop along quite different lines. The question of whether an atomic bomb would have been built, when, where and by whom, if the Second World War hadn't intervened, is not as hypothetical as it may seem.

It's fairly obvious that in a non-belligerent world, the development of nuclear energy for peaceful purposes would have preceded any military exploitation, as the innovations in cancer treatments, the use of X-rays and advances in technological applications had already shown. Or the first military applications, especially of reactor technology, might have been limited to small-scale projects, like nuclear propulsion systems for submarines. Perhaps, too, the use of nuclear technology for defensive purposes might have been more evenly spread between countries. Given the science, there can be absolutely no doubt but that at some point in the future, atomic bombs would be built.

Scientists, like Szilard or Einstein or latterly even Bohr, might convince themselves that an allied atomic weapons programme was about winning a race for the bomb against the murderous fascist regime of Hitler. Presidents and prime ministers would take a more long-term view: possession of the ultimate weapon of mass destruction was about bigger issues of geopolitics; a statement, as well as a guarantee, of political power and regional hegemony, irrespective of who else might be in the race to build an atomic bomb or might be tempted to join it in the future.

In the US, as many years later in Britain, the programme to build an atomic bomb came about through an incremental process rather than as the result of any single decision. On foot of Einstein's letter, Roosevelt had established the Briggs Committee to study uranium chain reactions. Although Szilard, Fermi and Wigner were members of the committee, it progressed much too slowly for Szilard's liking and again in 1940, he persuaded Einstein to sign letters addressed to Roosevelt and the committee chairman exhorting speedier action. At White House level, and particularly after the Americans became aware of British feasibility studies on the construction of an atomic weapon, the longer-term strategic importance of a weapon of such potential was fully appreciated.

'It is important therefore,' one report stated, 'that we gain the lead in this [atomic] development. The nation which first produces and controls the process will have an advantage which will grow as its applications multiply.'[7]

The Americans were not alone in this view. Joliot's 1939 papers on neutron emissions and Bohr's article on the nuclear force had attracted attention in Berlin, kick-starting a Nazi interest in developing a weapons programme. Responsibility for directing German fission research was promptly located within the new War Office of the Reich and on 19 September the German physicists were summoned to a conference in Berlin, the first of several, to consider the possibility of creating an atomic bomb. Germany's atomic bomb research programme was only finally abandoned in 1944.[8] The Imperial Army Air Force of Japan authorised research towards the development of an atomic bomb in April 1941, though that too would prove fruitless. A year later, in April 1942, the Russian physicist Georgi Flerov wrote to Stalin suggesting a Soviet research programme. Flerov had been alarmed by the disappearance from the scientific literature of any research findings on nuclear fission by Western scientists since 1939. He drew the logical conclusion: an official policy of prohibition on the publication of the results of experiments on nuclear fission existed in the West.

But it was in Britain, by mid-1941, that the greatest strides had been made towards developing what Otto Frisch described as the ultimate 'weapon of mass destruction'.

* * *

Before war came to neutral Denmark, Lise Meitner's nephew, Otto Frisch, was based in Copenhagen. Rightly fearing for his life and safety if Denmark was annexed by Germany, he relocated to Birmingham University and the Department of Physics run by the Australian Marc Oliphant in the summer of 1939. Frisch developed a friendship with the German Rudolf Peierls, who had arrived in England on a scholarship in 1933 and elected to stay put following the Nazi purge of the German universities. With the outbreak of war, both Peierls and Frisch were initially classified as 'enemy aliens'. Strictly speaking, as foreign nationals Peierls and Frisch could not engage in war work. In practice, they were busy working out how to build an atomic bomb.

Up to late 1939, all calculations based on Joliot's last published experiments suggested that to achieve critical mass, or the smallest volume of material required for explosive capacity, a uranium bomb would weigh tens of tons; far too heavy for any aircraft.[9] Otto Frisch reconsidered the options, in particular fast-neutron fission of uranium 235, while Peierls worked out the mathematics. To their astonishment, their calculations showed that, providing uranium 235 could be separated in sufficient quantity from ordinary uranium, two or three pounds of U235, rather than several tons, should be enough for an explosive device. Furthermore, its destructive potential was enormous.

In early 1940, the scientists brought their calculations to Oliphant, who advised them to write them out in a paper that became known as the Frisch–Peierls memorandum. Years later, Frisch would ask himself why, at that point, he didn't abandon the project and say nothing to anybody.

'Why start on a project,' he wrote 'which, if it was successful, would end with the production of a weapon of unparalleled violence; a weapon of mass destruction such as the world had never seen?' He answered his own question: 'We were at war and the idea was reasonably obvious; very probably some German scientists had had the same idea and were working on it.'[10]

The Frisch–Peierls memorandum provided a technical explanation of how the super-bomb should be constructed and how it would work. Even in the technical exposé, the scientists emphasised the dangers of radioactive fall-out.

'Any estimates of the effects of this radiation on human beings must be rather uncertain because it is difficult to tell what will happen to the radioactive material after the explosion,' the memorandum stated.

'Most of it will probably be blown into the air and carried away by the wind. This cloud of radioactive material will kill everybody within a strip estimated to be several miles long. If it rained the danger would be even worse because the active material would be carried down to the ground and stick to it, and persons entering the contaminated area would be subjected to dangerous radiations even after days.'

'Effective protection is hardly possible,' the scientists noted.

'The irradiation is not felt until hours later when it may become too late. Therefore it would be very important to have an organisation which determines the exact extent of the danger area, by means of ionisation measurements, so that people can be warned from entering it.'

In a covering note, Marc Oliphant also stressed the radiation dangers:

> Owing to the spreading of radioactive substances with the wind, the bomb could probably not be used without killing a large number of civilians,' he said. 'This may make it unsuitable as a weapon for use by this country.

A new doctrine of deterrence would override such moral qualms, Oliphant considered: 'If one works on the assumption that Germany is, or will be, in the possession of this weapon ... The most effective reply would be a counter threat with a similar weapon.'[11]

* * *

Joseph Rotblat managed to get on one of the last trains from Warsaw the day before the German invasion of Poland.[12] He was forced to leave behind his wife, Tola, who was ill with appendicitis. Rotblat had paid a flying visit to Poland to see his wife and bring her to England, where he had been working as a research assistant to James Chadwick, now Professor of Physics at Liverpool University. A few days after war was declared on Germany Rotblat drew Chadwick aside. 'We should start work on the bomb,' he suggested.[13]

Already, Rotblat had calculated how the uranium-triggering mechanism for an atomic bomb might work. To create a weapon of massive explosive power, all that was needed was a few pounds of highly fissionable material, such as uranium 235. Separated into two portions, held apart by a spring – a uranium gun – the element would spontaneously begin a self sustaining chain reaction if it was suddenly forced together, resulting in a devastating explosion.

On this last visit to pre-war Poland, Rotblat showed his calculations to his old professor, Ludwik Wertenstein, himself formerly a pupil of Marie Curie and then Director of the Radiological Laboratory of Warsaw. Wertenstein did not quibble with Rotblat's figures. He simply pointed out that he himself would never work on atomic weapons.

Rotblat never saw his wife again. Despite several attempts to rescue her from Warsaw, by arranging visas and transit routes through Italy and Denmark, the couple lost contact in 1940 and it is likely she became an early victim of the Holocaust.

* * *

Oliphant had forwarded the Frisch–Peierls memorandum to Henry Tizard, the crown official in charge of the scientific committee for Britain's air defence.[14] Just as Denmark and Norway were undergoing invasion, Tizard appointed a sub-committee, chaired by Sir George Thomson.[15] The committee included Oliphant, Chadwick and John Cockcroft. Frisch and Peierls were dismayed to find themselves excluded because of their alien status; but were eventually mollified when they, and Rotblat, were appointed as consultants.

Niels Bohr was visiting Norway when his country was invaded. Shortly after his return to Copenhagen, he and his wife Margarethe met briefly with Lise Meitner. On her own safe return to Sweden, Meitner sent a telegram to a friend in England to let him know the Bohrs were well, if unhappy about current political events. 'Please inform Cockcroft and Maud Ray Kent,' the telegram requested. John Cockcroft read into this an anagrammatic message that the Germans had seized radium supplies in Copenhagen. As it transpired, Maud Ray had been a governess to the Bohr family and lived in Kent. Unwittingly, she lent her name to Thomson's committee, duly renamed the MAUD Committee in June 1940.

The so-called 'phoney war' was about to come to an end. In May 1940, as the German tanks rolled over Belgium and Holland, Winston Churchill was appointed Prime Minister, replacing Chamberlain. The

British Labour Party, under Clement Attlee, had refused to participate in a wartime unity government under Chamberlain. Lord Halifax, one of the architects of the appeasement strategy and who might probably have favoured cutting a deal with the Germans, was the establishment choice to replace Chamberlain, especially at Whitehall. But Churchill, whose first speech as Prime Minister to a subdued House of Commons promised only 'blood, tears, toil and sweat', was the more obviously popular choice.[16]

Churchill described his aim as victory: 'Victory at all costs, victory in spite of all terror, victory however long and hard the road may be; for without victory, there is no survival.' The pious hope in such rhetoric drifted further apart from reality as the year moved on. First, there was the disastrous entrapment of the British and French armies at Dunkirk, and their heroic evacuation. Next came Operation Sea Lion and the Battle of Britain. As unacceptable losses mounted for the German Luftwaffe, Hitler personally ordered the bombardment of London. Up to the end of 1941, more than 20,000 Londoners lost their lives to the Blitz and a further 23,000 civilians were killed in other cities. The outcome of the war still hung precariously in the balance.

5

'Fat Man' and 'Little Boy'

> 'In most ruined cities you can bury the dead, clean up the rubble, rebuild the houses and have a living city again. One feels that is not so here.'
>
> US naval officer in a letter home to his wife from Hiroshima, September 1945

Against a backdrop of uncertainty, even about survival, the MAUD Committee made its first report in March 1941. A uranium bomb was 'likely to lead to decisive results in the war,' the committee concluded.

Their report supported the development and use of an atomic weapon:

> The points which we regard as of overwhelming importance are the concentrated destruction which it would produce, the large moral effect, and the saving in air effort the use of this substance would allow, as compared with bombing with ordinary explosives.

The MAUD Committee had commissioned a wide range of research projects, including work by the refugee German physicist Franz Simon at Oxford on the feasibility of using gaseous diffusion to separate uranium 235 from uranium 238. Simon proved the method, estimating the cost of producing one kilogram of uranium 235 at about £5 million sterling.

MAUD also consulted engineering experts at Cockcroft's old firm, Metro Vickers, and at ICI. They advised that uranium bombs could be produced within a two-year timescale, possibly by the end of 1943.

The committee recommended work 'be continued on the highest priority and on the increasing scale necessary to obtain the weapon in the shortest possible time', preferably in collaboration with America. 'Even if the war should end before the bombs are ready the effort would not be wasted, except in the unlikely event of complete disarmament, since no nation would care to risk being caught without a weapon of such decisive possibilities,' it observed.[1]

MAUD's second report in July 1941[2] explored the means of production, specifically reactors, for various isotopes that could be used in the manufacture of weapons. The refugee French scientists had been lodged at the Cavendish, where they had continued their research on slow-neutron bombardment of uranium 238. This had resulted by 1941 in the identification of two new elements with atomic numbers of 93 and 94 respectively. Amongst its other properties, 94 had a long half-life. It was also highly fissionable. Using a cyclotron in the laboratories at Berkeley, California, the American Glenn Seaborg, also in 1941, had isolated this new man-made element, which he named plutonium.[3]

* * *

Since mid-1940, America and Britain had exchanged scientific information, mostly concerned with progress on radar but also including some atomic material. As well as advising that Britain should establish its own programme, Churchill's wartime scientific advisor, the physicist Sir Frederick Lindemann, later Lord Cherwell, arranged for the MAUD reports to be passed to the United States.[4]

Initially, the Americans were slow to react. Vannevar Bush, chief scientific advisor to President Roosevelt and Chair of the US National Defense Council, and his deputy, James Conant, were sufficiently impressed by the progress the British had made to begin streamlining American research efforts. The MAUD reports became available to Bush and on 9 October 1941, he made a presentation of their contents to the President. By comparison with America, British efforts were focused and had clear objectives. Roosevelt gave Bush his seal of approval to an intensive programme to build atomic weapons. Like Churchill, he had reserved the right to take that decision himself.[5]

At last, the US-based scientists would secure the funding they had long sought to take their various projects beyond the desktop laboratory stage. But the end of their frustration came at a price: control was passing to the administrators and the politicians.

* * *

The British atomic weapons programme, code-named 'Tube Alloys' by Churchill, was highly classified – any knowledge of it was limited to a small group of ministers and advisers. Labour members of the cabinet, including Attlee, the Deputy Prime Minister, and Ernest Bevin, who was also a member of the war cabinet, were entirely excluded. Secrecy was also a hallmark of the American programme, ultimate authority for which was vested in the President. Most of the relevant congressional committees were by-passed; nor was the Vice-President informed in any detail about the programme.

Roosevelt wrote to Churchill in October 1941 suggesting full integration of the US and British bomb programmes, naturally under overall American control. In a foretaste of post-war Anglo-American relations on the atomic issue, misunderstandings and mistrust abounded at the level of officials and administrators. The Americans feared the British would later seek to commercially exploit knowledge gained at American expense during the wartime programme. The British were reluctant to hand over control of atomic developments to the Americans. The net result was that by mid-1942 all co-operation with the British had effectively ceased, while the US programme was rolling forward.

Towards the end of that year, on 2 December 1942, Enrico Fermi was preparing to carry out the experiment on a small graphite reactor at Chicago University, which, if successful, would put a final seal on the feasibility of building an American atomic bomb. Fermi's experiment induced a sustained chain reaction that lasted four and half minutes.

As Eugene Wigner described it:

> Even though we had anticipated the success of the experiment, its accomplishment had a deep impact on us ... We felt as, I presume, everyone feels who has done something that he knows will have very far-reaching consequences which he cannot foresee.[6]

* * *

Rated a highly classified military operation, the Manhattan Project to build the bomb was placed under the command of US Army General Leslie Groves.[7] Vast and wide ranging, the US programme included the construction of whole new towns and hundreds of miles of roadway to serve them, such as Hanford in Washington State, where large water-cooled graphite reactors were constructed to manufacture plutonium, and the complex at Los Alamos in the New Mexico desert, where scientists refined their calculations on how to make the bomb work. The first uranium bomb test, Trinity, was carried out in July 1945. The programme ultimately employed over 100,000 people and cost some $2 billion.

To Leo Szilard, the project might have seemed like a fantasy that had finally come true. But it was no longer driven by the scientists, nor, despite Szilard's idealism, could it be led by them. When their technical input was complete, their only remaining role would be as ineffectual Greek chorus to an unfolding tragedy.

By late 1942, Britain was experiencing problems with resources for Tube Alloys. In Churchill's own words:

> Great Britain at this period was fully extended in war production and we could not afford such grave interference with the current munitions programmes on which our warlike operations depended. Moreover, Great Britain was within easy range of German bombers, and the risk of raiders from the sea or air could not be ignored.[8]

Meeting in Quebec in August 1943, Churchill and Roosevelt signed off an executive agreement to integrate the British and American bomb projects. In this secret agreement they pledged, understandably, not to use atomic weapons on each other and secondly, not to use the bomb on any third party without each other's prior consent.

The leaders also pledged secrecy: 'We will not either of us communicate any information about Tube Alloys to third parties except by mutual consent.' In a clause that would lead to endless wrangling and disputes after the war, Quebec granted the US President a veto over future British exploitation of 'any post-war advantages of an industrial or commercial character'. Finally, the Quebec Agreement established administrative structures, representative of Britain, the US and Canada, to oversee the project and ensure raw material supplies, especially uranium.

With America now calling the tune, teams of British scientists moved to the US and Canada to take their places at the various centres of the Manhattan Project that sprawled throughout the United States and into Canada. James Chadwick became the Chief British Scientist at Los Alamos. Marc Oliphant eventually teamed up with his friend, Ernest Lawrence, at Berkeley to work on uranium 235 refinement. John Cockcroft, in 1944, was appointed Director of the Montreal and Chalk River Laboratories in Canada.

Labelling them communists and potential fifth columnists, General Groves rejected any notion of the French scientists' involvement at Los Alamos or any other projects on US soil directly related to bomb production. As a result, the Paris group found themselves in Canada working on reactor and isotope development with Cockcroft, who eventually persuaded the Americans to integrate the Canadian reactor experiments into the Manhattan Project.

Avoiding Groves' immediate supervision, they were luckier than they may have thought. Many of the scientists based in the US, including Fermi, Szilard and Oliphant, bristled at the general's security controls, particularly his insistence on compartmentalising the scientific aspects of the project. Nobody, except Groves himself and his hand-picked overall Scientific Director of the Manhattan Project, Robert Oppenheimer, was allowed to know what anyone else was doing.

Groves' attempts to impose a military regime on the scientists, including a proposal that all the scientists should be required to wear uniforms, were nearly enough to incite rebellion several times over. Szilard's particularly vociferous objections to Groves' strictures almost resulted in his arrest and incarceration at one point, and Groves had the Hungarian placed under surveillance as a possible spy. He was, Groves remarked, 'the kind of man that any employer would have fired as a troublemaker'.[9]

The net effect of this obsession with secrecy and security, however, was that it worked brilliantly, but only with regard to the general public in America and Britain and the governments and peoples of other countries friendly to the Allied cause. As events transpired, the policy was far less effective in denying information to what Groves had identified from early on as its primary target – the Soviet Union.

* * *

Every time he looked at the map of the western approaches that hung on the wall of his office, Churchill, brought back into the British cabinet at the start of the war as First Lord of the Admiralty, was reminded of 'the numbing loss of the Southern Irish ports'.[10]

De Valera had not waited for the formal declaration of war against Germany before hastily reconvening the Dáil on 2 September 1939 from its summer recess to officially announce Irish neutrality. The continuing partition of Ireland made it impossible for Ireland to adopt any stance other than neutrality, De Valera said. 'We, of all nations, know what force used by a stronger nation against a weaker one means,' he told the Dáil.¹¹ 'We have known what invasion and partition mean; we are not forgetful of our own history and, as long as our own country, or any part of it, is subject to force, the application of force, by a stronger nation, it is only natural that our people, whatever sympathies they might have in a conflict like the present, should look at their own country first and should, accordingly, in looking at their own country, consider what its interests should be and what its interests are.'

De Valera himself had no doubts about Ireland's interests. In so far as Ireland was still a member of the Commonwealth of Nations – a vague matter on which the British and Irish held separate and opposing views – Ireland was the only Commonwealth state that failed to side with Britain at the outbreak of the war. By 10 September, De Valera had furnished the British government with a memorandum denying Britain the use of the treaty ports, a policy that particularly incensed Churchill, who went well beyond his admiralty brief in a series of bellicose demands to his own colleagues that the Irish be forced to concede them.¹²

In the days following the declaration of war, and for the first time since its foundation, the Irish state found itself on the receiving end of a large flow of immigrants from Britain, many of whom were Irish workers in Britain and their families who opted to sit out the war in the presumed safety of a neutral Ireland. Before long, however, the tide was once again reversed as the rural and urban poor who needed jobs wherever they could find them, those whose families had a tradition of fighting in the British forces and those who believed in their hearts that the coming war against fascism in Europe was right and just, about 100,000 in all, began the trek to Britain to work in British munitions factories or to join the British army, navy and air force.¹³ In one of the more curious twists of the wartime relationship between the two countries, De Valera requested Britain not to allow Irishmen returning on leave from the British forces to wear army uniforms home. Throughout the war, Britain obligingly maintained dumps of civilian clothing at Holyhead to meet this requirement.¹⁴

However much it infuriated Churchill, not just because of the loss of the ports to Britain but also from an imperialist perspective of betrayal, as he saw it, by a lesser dominion of dubious legal status, De Valera's policy of neutrality was firmly rooted in domestic political pressures. Aware that elements among the IRA were in contact with Germany and fomenting alliances to pursue their own campaign against Britain, equally aware that any declaration to take Britain's side might fatally expose fissures within his own cabinet, De Valera opted for the political middle ground of neutrality.[15]

The threat of a renewed civil war, or massive civil unrest that might follow from siding with Britain, may or may not have been real. As it was, De Valera was not prepared to take a chance on it. He had strongly favoured appeasement and supported Chamberlain's efforts to maintain peace in Europe. He was bitterly disillusioned by the failure of the League of Nations, of which he had been President in the 1930s, to maintain international peace and curb the worst instincts of the fascist regimes in Europe. At the outbreak of war no one could reliably predict which side would ultimately prevail. In such circumstances, De Valera believed a small independent country should be minded to look after itself. In any case, Ireland had neither the armaments nor the resources to adequately defend itself in the event of any external military threat.

Neutrality was popular in the Free State; and the onset of the war further relieved De Valera and his party, Fianna Fáil, of growing domestic criticism about the failure of its economic and social policies.[16] Ireland would experience an 'Emergency', as opposed to a war, implemented via a rigid and uncompromising policy of media and political censorship. From De Valera's point of view, suppression within Ireland of partisan views favouring one side or the other in the war was important to the maintenance of stability within the country. The partition of Ireland provided a convenient hook on which to hang this policy and, if only in his singular view, provide international justification for it.[17]

Three days after the fall of Paris, the British government dispatched Malcolm MacDonald to Dublin to meet with De Valera. In Britain's greatest hour of need, MacDonald was charged with persuading De Valera to enter the war on Britain's side and abandon neutrality. Over the next three months, an extraordinary series of discussions took place within and between the two governments, and offers and counter-offers were tabled. The gist of it all was that Britain offered De Valera a united Ireland after the war in exchange for the use of the ports and a number of troop bases.

Final rejection of the proposals came from De Valera in July 1940. The offer of unity was 'tentative', De Valera wrote on behalf of the Irish government:

The plan would involve our entry into the war. This is a course for which we could not accept responsibility. Our people would be quite unprepared for it, and Dáil Éireann would certainly reject it.[18]

Whatever else, the Northern unionists, under Lord Craigavon, would have resisted any such plan, or possibly violently split apart over it.[19] Further, if Britain lost the war, as appeared likely at the time, an indefensible Ireland would be easily occupied by Germany.[20] Even if Britain won, there was no guarantee that the wartime promise could be honoured, without Ireland becoming engulfed, yet again, this time possibly in a sectarian civil war, over the old issue of unification.

In making his appeals to De Valera, MacDonald pointed out time and again that such an opportunity for a united Ireland 'might never recur'. In this, MacDonald would prove prophetically correct. In the meantime an uneasy relationship prevailed between Britain and Ireland, particularly on the issue of wartime trade and securing arms for the defence of the Free State.

The Irish government had made little preparation for the impact of the war on Ireland, either to defend the country against invasion or secure imports of raw materials for industry or food for its population. As naval losses mounted in the Atlantic, Churchill's anger at the loss of the Irish ports to Britain grew in intensity. In a last-ditch effort to secure use of the ports for Britain, in June 1940 Churchill telegraphed Roosevelt. 'We are also worried about Ireland,' he wrote. 'An American squadron at Berehaven would do no end of good I am sure.' No American naval visit took place.[21]

The British arbitrarily cut quotas of raw materials supplied to Ireland, but still held the Irish to the terms of the 1938 Trade Agreement on sourcing supplies; part of a deliberate policy by Churchill to both punish Ireland for its neutrality and exert pressure on De Valera to reconsider. War rations in the Free State of some commodities, such as tea, of which the Irish were notoriously fond, dwindled to only a quarter of the monthly allowance in Britain. Personal travel by car was banned for much of the war because of an acute shortage of petrol. By 1941, the government had established the Irish Shipping Company in an effort to secure imports into Ireland under its own steam.

When America entered the war in 1941, Churchill again invited De Valera to join the Allies in an effusive telegram that De Valera mistook for yet another offer of post-war unification of Ireland. Churchill wrote: 'Now is your chance. Now or never. "A Nation Once Again". Am ready to meet at any time.'[22]

Arguably, all he meant to imply was that Ireland, by coming into the war, could lay claim to some measure of moral redemption. Irrespective of what Churchill intended, De Valera was determined to stick fast to Ireland's policy of neutrality. Churchill rapidly made it known that if Ireland remained neutral to the end of the war, ending partition would be off the table for several generations to come.

But America's entry into the European war and its likely influence on the final outcome moderated De Valera's external implementation of his neutrality policy. From now on, Irish neutrality assumed a more benevolent aspect, tilted towards the Allies: the crews of Allied planes crash landing in the Free State were quickly moved across the border to Northern Ireland and no longer interned.

To the British, De Valera consistently presented the Irish public's overwhelming endorsement of neutrality as an insuperable political difficulty. De Valera could point to the opposition leaders who marched at the head of massive public rallies in support of neutrality in Dublin in 1940. The reality was a little more complex. Stultifying censorship blacked out practically all media references to what was happening in the war, including any references to Nazi atrocities. Ordinary Irish citizens were deprived of sufficient information on which to make their own judgement of events. Even death notices of Irish soldiers killed on the Allied side were rigorously censored. Frank Aiken, as the minister responsible for the government's censorship policy, threatened to shut down the *Irish Times*. Throughout the war, the editor of the *Irish Times*, Robert Smylie, who described Aiken as 'unintelligently impossible', expended much time and energy seeking to outwit the censorship regulations.[23]

De Valera's fear that Ireland might be invaded by the Germans, or the Allies, at some stage in the conflict was not wholly without justification.[24] Apart from keeping the country from starving, and maintaining the best defensive capacity possible, maintaining the Radio Telefís Éireann transmitter at Athlone was a key national priority. RTE would provide the only means of mass communication by government to the people if the country was invaded by either side.

Ernest Walton, who averaged '17 hours a week' teaching in an understaffed physics department at Trinity throughout the war years, found much of his spare time taken up with making high-power valves to keep the RTE transmitters going as well as locating metal and spare parts to repair hospital X-ray machines. Metal of any kind was in extremely short supply. The authorities suggested Walton make do with sheets of metal harvested from German aircraft that had crashed in the Free State.

C.P. Snow, the physicist and author whom Walton had known at the Cavendish, had contacted him early in the war and attempted to recruit him to work on radar detection projects. Walton's professor at Trinity, R.W. Ditchburn, had already left for the UK to engage in radar research at the start of the war. Not surprisingly, the Provost of Trinity was reluctant to allow Walton to follow suit. Walton later received two further offers, from Chadwick and Oliphant, asking him to join scientists going out to America to work on projects that remained unspecified, but were obviously related to the development of the atomic bomb. But his response remained the same: his duty was to remain in Ireland and at Trinity where he was needed.

In a political context, De Valera too had fulfilled his duty in preserving Irish neutrality, and observing it to the letter, throughout the war years. He could claim the support of the broad mass of the Irish people for his policy. By his own logic, neutrality internationally signposted Ireland's independence more effectively than any other measure he had taken since coming to power. In March 1944, De Valera refused to expel Axis diplomats from Ireland and close their legations, despite American pressure in advance of D-Day. His personal decision, against the advice of his own civil servants, on the afternoon of 30 April to visit the German ambassador, Hempel, to pay his respects on the death of Hitler on behalf of the Irish people outraged public and political opinion in Britain and the US.[25] In the fevered and traumatic atmosphere that prevailed at the end of the European war, the protocols of neutrality that obliged De Valera to present his condolences for Hitler were not much appreciated.

* * *

On 4 July 1945, the day before the British general election that would result in a landslide victory for Clement Attlee's Labour Party, Britain gave its approval to the United States to use the atomic bomb on Japan. Franklin D. Roosevelt had died on 12 April 1945, some three weeks before Hitler's suicide in the Berlin bunker and Germany's final collapse. The war in Europe ended with Germany's unconditional surrender on 7 May, but from early spring the US administration had been actively considering a list of Japanese city targets for the atomic bomb.[26]

In his few months as Roosevelt's Vice-President, Truman, had been told practically nothing of the details of the Manhattan Project. In his earlier Senate career he was actively discouraged from pursuing any information related to it. Now President Truman was confronted with the final decision on using atomic weapons to force an unconditional surrender from the Japanese and end the Asian war.

There were other compelling reasons to use the new ultimate weapon of war, not least the recognition by both the United States and Britain that once the war was over, the alliance with Soviet Russia would dissipate into mistrust if not confrontation. Following the successful Trinity test of the uranium bomb in New Mexico, Truman had elected to drop a subtle hint to Stalin at the Potsdam Conference in Germany. Contemporary American observers were surprised by the Soviet leader's lack of reaction. In fact, Stalin already knew of the bomb's development from his spies at Los Alamos. Within the Soviet Union itself, work on the theoretical side of atomic weapons development was well advanced.

Apart from laying claim to a new world order, as possession of the bomb was meant to signify to the Russians, or bringing the Japanese war to an end, there was a further issue influencing the decision on whether or not to use the new weapon: cost. Sooner or later the war would end and congressional committees would ask questions about the disposition of $2bn to make a weapon.[27] The only retrospective justification for such expenditure, however barbarous, was to demonstrate its effectiveness.

* * *

On 6 and 9 August 1945, respectively, atomic bombs were dropped on the Japanese cities of Hiroshima and Nagasaki. The uranium bomb, 'Little Boy', weighing four tons, the equivalent of 12,500 tons of TNT, was dropped on Hiroshima and exploded at a height of some 1,500 feet above the city. Of Hiroshima's 76,000 buildings, 70,000 were damaged or totally destroyed. Tens of thousands of its inhabitants were killed in the immediate explosion. Injury, mainly severe burns and radiation sickness, would raise the death toll to 130,000 by the end of the year. 'Fat Man', the plutonium bomb that exploded 2,000 feet over Nagasaki a few days later, claimed over 70,000 lives by the end of 1945. Over the next five years, the death toll in Hiroshima attributable to the atomic explosion increased to 200,000 and in Nagasaki to 140,000.

Most chilling of all was the death rate among bomb casualties for the rest of 1945; 54 per cent in both cities.[28] This compares with a death rate of 10 per cent among casualties of the firebombing of Tokyo in March 1945. Nuclear weapons wrought total destruction, wiping out all the vestiges of human civilisation along with human life itself.

Settled down to lunch with a group of sailors aboard the *Augusta*, on his way back from the Potsdam Conference, Truman was informed of the successful mission over Hiroshima. 'This is the greatest thing in history,' he remarked. In Chicago, writing a letter to a friend, a disillusioned Leo Szilard described it as 'one of the greatest blunders of history'.

PART THREE

NUCLEAR PROLIFERATION

6

Britain's Bomb Factory

'We were assured that when the war was over these explosive factories would be demolished and desecration of the coastline would cease. Now we are threatened with some new sort of devilry!'

West Cumbrian resident, 1947

At the end of the Second World War, it would have been hard to imagine that Whitehaven, on the Cumbrian coast, was once England's third largest trading port. Whitehaven's days of glory dated back to the seventeenth century when Sir John Lowther, Earl of Lonsdale, first laid out a plan for the town, reputedly modelled on Sir Christopher Wren's plans for the reconstruction of London after the Great Fire. The town Lowther built, with its handsome Georgian facades and neat elegant squares, had previously been a fishing village of about six houses.[1]

Whitehaven's prosperity was built on its coal trade with Ireland. The celebrated eighteenth-century Dean of St Patrick's Cathedral, Jonathan Swift, spent his early childhood in Whitehaven. Local legend dubiously insists its bustling port later provided the inspiration for Lilliput in *Gulliver's Travels*. Swift's own account was more prosaic. Shortly after he was born, his nurse, a native of Whitehaven, was called back to England from Dublin to attend to a sick relative from whom she hoped to inherit a legacy. Too fond of the baby to leave him behind, she stole him away on the ship to Whitehaven. When Swift's mother discovered what had happened, she insisted the child remain where he was as he was too sickly to make a return journey. Swift claimed that by the time he was brought back to Dublin in 1671, aged four, he could read from any chapter of the Bible.

From the 1660s, the Lowther family exploited West Cumberland's extensive coal reserves to supply Dublin and other Irish ports with coal. On the back of the coal trade Whitehaven developed a ship-building industry. The double-bottomed ships used for the coal runs were robust enough for transatlantic voyages and the Whitehaven traders transported Irish goods to the colonies of Virginia and Maryland, returning with supplies of tobacco for Ireland, Scotland and northern England. Whitehaven engaged in the African slave trade throughout the eighteenth century. Slavers ran to the Caribbean exchanging their human cargo for rum.[2] Whitehaven also held the distinction of being the only English mainland port bombarded by American ships, commanded by John Paul Jones, during the American War of Independence.

For over a hundred years, West Cumberland's thriving industrial landscape attracted workers and their families from across northern England and Scotland and along Ireland's east coast as well as its northern counties. Deposits of rich haematite iron ore, 60 per cent pure and free of phosphorous, occurred abundantly and were mined to provide essential raw material for the British steel industry. Woodland was planted to supply charcoal. Mills and foundries sprang up throughout the county.

Industrial prosperity was short-lived. The Solway Firth was too shallow for steamships, and by the start of the nineteenth century Whitehaven was losing trade to Liverpool. The larger coalmines located directly along the coastline and under the sea had always been especially dangerous due to the structural impossibility of venting off pockets of explosive gas. The death rate from mining accidents was high, even by the safety standards of the time. Most of the main coal seams were exhausted by the early 1900s. In 1901, there were forty-three collieries; by 1930, the number had shrunk to twenty-eight. The haematite iron ore reserves were worked out by the 1920s.

A 1934 Department of Labour survey described the village of Cleator Moor, five miles south of Whitehaven and often referred to locally as 'Little Ireland' because of its predominantly ethnic Irish population, as 'both drab and depressing'. Cleator Moor and Whitehaven experienced sectarian tensions, especially marked in the late nineteenth century, involving violent clashes between the Irish ore miners of Cleator Moor and the Orange lodges of Whitehaven. Finally, in the 1920s the local council banned Orange parades in Cleator Moor following a spate of riots and public disorder.

'The town is almost wholly composed of workmen's dwellings of a type which must have presented a not very cheerful aspect even in times when industry was thriving,'[3] the 1934 report said.

A few miles west of Cleator Moor, the sea coast townland of Drigg was selected by the Ministry of Supply as a site for a Royal Ordnance factory to manufacture TNT in 1941. Two years later, another TNT factory went into production on the neighbouring Sellafield site. Surrounded by mountains at the top of England's Lake District and bound by the sea, the West Cumbrian munitions factories were suitably remote from the risk of raids by German bombers.

At the end of the war the munitions factories shut down. Local confidence received a boost with a Board of Trade announcement that Courtaulds, then a major textiles manufacturer, would create 2,000 new jobs on the Sellafield site. Courtaulds cancelled their factory plans for Sellafield in early 1947 when they became aware of the Ministry of Supply's proposal to build a plutonium production plant on the neighbouring Drigg site. Courtaulds cited local labour shortages as the reason for pulling out. The number of local workers available, the company claimed, was not sufficient to satisfy the labour demands of both projects.[4] Courtaulds sold the Sellafield site back to the Ministry of Supply.

As Hugh Gethin Davey, general manager of the Windscale Works from 1947 to 1958, later noted:

> In 1947, the horrors of Hiroshima and Nagasaki were still fresh in the public mind, and to the great majority of people the term 'atomic energy' was associated only with death and disaster. Probably a proposal to build an atomic energy establishment anywhere in the British Isles at that time would have met with some opposition from the local people and, in this respect, the people of West Cumberland were not exceptional ... To the average Cumbrian it appeared that a safe project, devoted to the manufacture of materials for peacetime uses, had been replaced by something connected with atomic bombs.[5]

* * *

Attlee's Labour government had taken over a Britain practically bankrupted by the war. Britain had lost a quarter of its wealth. At the most basic level its infrastructure was devastated, with more than one million homes in ruins. What remained of its empire was rapidly disintegrating and the high expectations of its people for social reform that had unexpectedly ejected Churchill and the Tories from office in the first post-war election looked set for disappointment. On the day Japan surrendered, the internationally acclaimed economist John Maynard Keynes warned the new Labour government that without immediate financial assistance from the US, Britain was 'virtually bankrupt and the economic basis for the hopes of the public non-existent'.[6]

In retrospect, Attlee has come to be regarded as one of Britain's greatest prime ministers of the twentieth century. As a prime minister, he regarded himself as a 'first, among equals', allowing considerable autonomy to his ministers and operating a consensual rather than presidential style of government. By the time he left office in 1951, the foundations of Britain's welfare state were in place. The National Health Service had been established and a range of industries, including coal, electricity and the railways, brought into public ownership. But when it came to foreign policy and defence, the difficulty for Attlee and his cabinet strongman, Foreign Secretary Ernest Bevin, lay not in taking on new ideas but in jettisoning old ones.

Attlee and Bevin were products of their time, imbued with Britain's imperialist ideology and tradition. Both were strongly anti-communist and suspicious of Soviet Russia's intentions, particularly in Europe. Both apparently found Britain's new post-war status as a 'great nation' rather than a 'great power' difficult to accept.[7] Committed to a radical programme of social and economic transformation of Britain, their foreign policy amounted to a continuation, sometimes even more hawkish, of the imperialist tradition of the previous half-century.

Attlee personally believed the atomic bomb made conventional methods of warfare redundant. In the immediate post-war period, he was drawn to a system for international control of atomic weapons, and half-heartedly pleaded the case for talks with the Russians, without ever mentioning them, in a letter to President Truman in September 1945.

'The successful manufacture of bombs from plutonium shows that the harnessing of atomic energy as a source of power cannot be achieved without the simultaneous production of material capable of being used in a bomb,' Attlee wrote to Truman. 'Statesmen of the Great Powers are faced with decisions vital not merely to the increase of human happiness but to the very survival of civilisation,' he argued, urging a 'fresh review of world policy and a new valuation of what are called national interests'.[8]

Attlee had sent a copy of his letter to Churchill, who threatened to criticise it in public if its contents ever came to light. As far as Churchill was concerned, it was not clear what the Prime Minister was asking the American President to do. Irrespective of that, a major gap in the Anglo-American interpretation of what post-war atomic co-operation should mean was already beginning to emerge: neither the British nor the Americans were capable of walking in one another's post-war shoes. From the US standpoint, the 'special relationship' – a phrase coined by Churchill – could never be a relationship of equals. The Second World War had transformed the US into the world's leading economic and military superpower. To the Americans, possession and control of the atomic bomb and its secrets were emblematic of their new superpower status. To the British, lack of access to technical and material data on the bomb simply underlined their relentless drift away from the era of imperial glory.

Attlee's letter led to talks in Washington in November 1945 between the erstwhile wartime partners, Britain, Canada and the United States. The public face of the talks, as set out in the subsequent Washington Declaration, was to 'prevent the use of atomic energy for destructive purposes' and promote 'peaceful and humanitarian ends'. The Western leaders proposed the establishment of a UN commission 'for control of atomic energy to the extent necessary to ensure its use only for peaceful purposes'.[9] Privately, the main agenda concentrated on revamping the secret wartime Quebec Agreement and the even more secretive 1944 pact between Roosevelt and Churchill, known as the Hyde Park Agreement, of which there was no official record, but which had promised post-war co-operation on industrial applications of nuclear technology.

Attlee and his officials came away from Washington believing they had secured American agreement to technical co-operation, admittedly more implied than explicit. In return, the Quebec Agreement clause requiring UK consent before atomic weapons could be used had been watered down to consultation.

The Americans latterly took an even more restrictive interpretation of the Washington Declaration and its various side-agreements. Technical co-operation remained off limits. Their priority, it turned out, was to wrest as much control as possible over the world supply of uranium by securing access to Britain's Belgian Congo stockpiles[10] on preferential terms. There was little enthusiasm in Washington for any post-war British atomic programme.

Within six months, Attlee was again writing to Truman, complaining about the US refusal to share technical information, especially on reactors, and British fears that information restrictions were about to get worse. He never received a reply, not even an acknowledgement.[11]

Events took a further twist with the unmasking in Canada of a spy ring among the Chalk Ridge scientists, involving the British physicist

Alan Nunn May. This was the first of a series of spy scandals involving British scientists and civil servants, and US public opinion was inflamed when the affair became public in early 1946. Nunn May had passed microscopic quantities of fissile materials to the Russians, in the interests, he claimed, of 'the safety of humanity'.[12] Nunn May's motives probably had more to do with self-preservation than saving the human race. Approached by the Russians in Canada in 1944 and asked to resume pre-war espionage activities, at first he refused, claiming he was now out of that business. His contact coldly advised him that Soviet spies were not permitted to resign.[13]

To protect their supposed atomic monopoly, and in complete ignorance of the 1943 Quebec Agreement and other secret undertakings on co-operation with Britain, the US Senate proposed the McMahon Act, which became law in October 1946. The US Atomic Energy Act established the structures to manage the development of atomic energy in the US – the US Atomic Energy Council (USAEC), which owned nuclear materials, and the Joint Committee on Atomic Energy (JCAE), comprising senators and congressmen and conducting its business in secret, to decide policy. The Act also imposed severe penalties, up to and including the death penalty, on sharing technical information either for military or industrial purposes, with parties outside the United States. British scientists had estimated that full US co-operation would cut development time on the atomic bomb project by at least a year.

* * *

In a telegram from the US in September 1945, Chadwick had set the ball rolling again on a British atomic project. A programme 'of our own,' he said, was essential to the defence of Britain and of the Commonwealth. The military chiefs of staff concurred, outlining, a month later, the deterrent value of atomic weapons. 'To delay production pending the outcome of negotiations regarding international control might well prove fatal to the security of the British Commonwealth,' they argued.[14]

Chadwick favoured building a plutonium production plant in Canada. John Cockcroft, in charge of the reactor research programme in Canada since 1943, agreed. The military preference was for a project based in Britain. It was a moot point. 'Virtually bankrupt' Britain didn't have the dollars to finance a Canadian-based project.

Rather than appoint a minister for atomic energy from within the ranks of his own party, Attlee acquiesced to civil service advice to co-opt Sir John Anderson, former Chancellor of the Exchequer in Churchill's wartime cabinet, in a quasi-ministerial role to head up an advisory committee. Anderson's virtue was that he had become fully acquainted with the atomic programme throughout the war years.[15] He was also known to have firmly opposed a post-war American monopoly on atomic weapons.

The aristocratic Scotsman, who in an earlier Foreign Office career had served as British Undersecretary to Ireland during the War of Independence, introduced the internationally infamous Black and Tans and Auxiliaries forces to the South of Ireland and approved the establishment of the notorious B Special auxiliary police force in the North, may have had all the right technical credentials to act as chief political adviser to the Labour government on atomic energy policy. But he was the almost complete opposite to Attlee in personal style as well as political philosophy. At the Washington meeting with the Americans in November 1945, General Groves resented what he regarded as Anderson's arrogant manner and made no effort to conceal his dislike.[16] Nor could Anderson restrain himself from indulging in regular public criticisms of the Labour government's welfare state agenda. Within two years he had resigned to take up a director's post with ICI.[17]

Attlee was also persuaded by the civil service mandarins not to house the programme within the specialist Department of Science and Industrial Research. Instead, responsibility was allocated to the larger Ministry of Supply, with its considerable wartime experience in setting up and managing ordnance factories around the country. Lord Portal, architect of the RAF strategic bombing campaign against German cities during the war, was made Controller of the Department of Atomic Energy in March 1946.[18]

In contrast to his otherwise open style of government, Attlee restricted information about the project to a select group of ministers who formed the General 75 cabinet sub-Committee. The Foreign Secretary, Ernest Bevin, towered over Gen 75, as he did over the cabinet generally; but the committee also included economics ministers – the Chancellor of the Exchequer, Hugh Dalton, and Stafford Cripps, then Minister for Economic Affairs. Given the country's financial straits, the economics ministers might have been expected to exercise some restraint over extravagance, although final decisions on atomic energy rested with the Prime Minister. Gen 75 made its recommendations to Attlee. Lord Portal also reported directly to the Prime Minister rather than the departmental head at the Ministry of Supply.

A decision to mount a research programme, based at Harwell, near Oxford, was taken in October 1945. John Cockcroft was recalled from Canada to head it up. Three months later, Gen 75 recommended construction of one plutonium production plant. On New Year's Day, 1946 the chiefs of staff reported to the Prime Minister a requirement for a stock of atomic bombs 'in hundreds rather than scores'.[19] One reactor became two, at a total estimated cost of about £35m.[20] An ICI engineer, Christopher Hinton, had already been recruited to take charge of the production programme, headquartered at Risley, near Warrington, then in Lancashire. William Penney, who had flown with the bomb crew over Nagasaki to measure the effects of the atomic blast and provide assessments on the post-war Bikini bomb tests for General Groves, was appointed director of bomb production in a temporary facility set up first at the military base of Fort Halstead and moved in 1950 to Aldermaston, near Cambridge. On 1 November 1946, Penney submitted a scheme for the ordnance section of an atomic bomb to Portal, typing the document himself to guarantee its absolute secrecy.

In October 1946, Gen 75 met to consider a proposal to build a gaseous diffusion plant to manufacture uranium 235. The plant had a development price tag of about £30m. Dalton and Cripps were against it for financial reasons. They were overruled by the Foreign Secretary, Ernest Bevin, who arrived late to the meeting claiming he had fallen asleep after a good lunch. Bevin was full of bile at the manner in which he felt Britain's atomic ambitions were being treated by the Americans and the way in which he had only recently been spoken to by the American Secretary of State, Martin Byrnes.

'We have got to have this thing over here whatever it costs,' he was reported to have said. 'And we've got to have the bloody Union Jack flying on top of it.'[21]

Through such a process of incremental decisions, Britain moved inexorably towards the construction of atomic weapons, although the Prime Minister refused to make any official commitment beyond research right up to the beginning of 1947.[22] Gen 75 then re-formed as Gen 163 – the same group of ministers minus their financial colleagues – and in 'the utmost secrecy' on 15 January 1947, made the decision to build an atomic bomb.[23]

Neither the cabinet nor the House of Commons was informed; nor Labour's scientific advisors. As late as July 1947, Sir Henry Tizard, making a presentation to the chiefs of staff on Britain's future defence requirements, was unaware that a political decision to build the bomb had already been taken.[24] Tizard had argued against Britain acquiring nuclear weapons on both financial and security grounds. Attlee ignored all such arguments. New financial protocols were devised to conceal expenditure on the atomic programme under diverse headings within the defence budget. Over the first four years of the project, total expenditure amounted to about £100m. Eighteen months later, in May 1948, a parliamentary question was planted with a desultory one-line answer that acknowledged existence of the atomic weapons programme, the first time it was formally revealed to the House of Commons.[25]

* * *

Nobody in Britain knew how to build a plutonium production plant. The Americans had kept foreign scientists, including the British, well away from the water-cooled Hanford reactors. For the purpose of making bombs, plutonium was much more efficient than uranium 235 – ten times as much uranium 235 as plutonium was required to produce half the TNT equivalent. The scientists calculated that an efficient plutonium production reactor could produce enough raw materials for fifteen bombs a year. Chadwick and Cockcroft also believed plutonium-based technology held more potential for the long-term commercial and industrial development of nuclear energy.[26]

Lord Portal, who admitted he knew very little about atomic energy, was intent on copying the American model of a water-cooled reactor. At first Portal resisted all advice from Cockcroft's team on any alternatives. He rejected Hinton's suggestion that one of the two proposed production plants should be gas-cooled – a dual-purpose reactor that would supply electricity as well as producing plutonium – on the grounds that military imperatives superseded any industrial applications of nuclear technology.

When Portal asked Groves for his advice about reactors at a meeting in the US in May 1946, the general replied that the best idea was not to build one at all.[27] The US rule of thumb for reactor location was fifty miles from the nearest town of 50,000 inhabitants; twenty-five miles from one of 25,000 people and five miles from any town with more than 1,000 inhabitants. The Hanford reactors had been located in a relatively unpopulated and isolated area in Washington State. A thirty-mile, four-lane highway was built to facilitate rapid evacuation of civilians as much as for the efficient transport of materials to and from the plant. Every morning, Groves said, he expected to wake up to the news that one of the Hanford piles had 'gone up'.[28]

In Britain, only Arisgair, a site at the most northern tip of Scotland, could meet the Hanford criteria for water supply and distance from population centres. Few people lived in the area, but the site lacked even the most basic infrastructure. Portal was eventually convinced that the Harwell design for an air-cooled reactor was not only technically feasible, but likely to be much safer in operation.

For a start, there was less risk of the reactor overheating if the water supply was cut off for any reason.[29] An air-cooled design would not require the thirty million gallons of water per day needed by the Hanford water-cooled piles. Because of its smaller dimensions, two or three reactors could be built, spreading the risk of failure and reducing overall construction costs by a third. Isolation from major population centres was still essential, but the air-cooled design opened up a broader range of possibilities, among them Harlech in Wales as well as the former ordnance factories in West Cumberland. Apart from its significance as a national heritage site, Hinton foresaw local community problems in Wales, not least bridging the language barrier in a Welsh-speaking area.

Sellafield was announced as the location for Britain's atomic energy site in July 1947. The site was renamed Windscale, after a bluff on the site, ostensibly to avoid any confusion with the uranium fuel plant at Springfields near Preston in Lancashire, itself the former location of a wartime poison gas plant. To cater for future expansion, the Ministry of Supply purchased an adjoining farm across the Calder river on the southern boundary of Sellafield, and leased it back short-term to the original owner.

Several of the old ordnance factory buildings – the administration block, workshops and canteen – were re-adapted for use. As well as the plutonium piles, a suite of chemical plants was erected to separate out the plutonium from irradiated fuel rods, manufacture plutonium metal and treat contaminated liquid wastes before discharging them to sea. To avoid cross-contamination and maintain security, the site was divided into four areas – the reactor or pile buildings, two separate sites for chemical processing and storage, and a common facilities area. Access of personnel and materials to each area was controlled by a pass system.

Construction of the Windscale piles began in September 1947. In October 1950, Pile No. 1 went critical, followed in June 1951 by Pile No. 2. West Cumbrians had become more reconciled to the 'atom bomb factory', as it was locally known, in their midst. Among other things, they had discovered their bargaining strength with central government. In an era of general cutbacks throughout Britain on education, Seascale, a coastal village six miles from Windscale, acquired a new primary school and Whitehaven a new technical college, and the site was providing employment to more than 1,000 local people.[30]

While it was generally accepted at a political and military level that there was no immediate risk of war, the programme for possession of the ultimate weapon of deterrence was characterised by urgency with little time even for laboratory-scale pilot plants to test nuclear or chemical processes. Even construction of the Windscale piles involved piecemeal innovation. Returning from a visit to the US Oak Ridge National Laboratory, late in 1948, John Cockcroft, much to the annoyance of engineers at Windscale, insisted on the installation of filters to capture particle emissions from the stacks.[31] The suggestion came too late to place the filters at the chimney base. Filter galleries were installed, less than ideally, at the top, creating an incongruous square overhang on the 125-metre Windscale stacks.

Windscale's General Manager, Hugh Gethin Davey, later described the frenetic pace of production that culminated in the delivery of sufficient plutonium for Britain's first atomic bomb test, codenamed Hurricane, on a ship off the Monte Bello Islands in the Indian Ocean on 3 October 1952, as a relay race that ended 'with a blinding, mushroom-shaped cloud of fire and a seismic explosion'. The first stage of the race, he noted, began with the production of the first plutonium at Windscale on 25 February that year.

'It would be futile and hypocritical to pretend that 25 February was just another day,' Davey wrote. 'Men who had spent the greater part of the dark silent hours with mental pictures of vessels, columns and interconnecting pipe work and intensely radioactive liquors flowing into, out of, and through them, could not awake to just another day. Ordinary days do not bring tensions almost unendurable, or give rise to lurking fears and desperate hopes.'[32]

* * *

The first Russian bomb test, 'Little Joe', detected by a joint US–UK project in 1949, gave added impetus to the British project.[33] The received wisdom until then was that Russia would not be able to develop atomic weapons to testing stage until the mid-1950s. A standard insider Washington joke in the mid-1940s was that the Russians couldn't deliver a bomb in a suitcase, because they didn't know how to a make a suitcase. Leading physicists, such as Hans

Bethe, had warned of a much shorter timescale of three to four years: Russian physicists were expert in nuclear theory, nor was there any lack of competent Soviet engineers.[34]

The Soviet weapons programme may have occupied a low political priority status throughout a war for survival against the Nazis and the hideous atrocities surrounding it, but the US demonstration of the power of atomic weapons at Hiroshima provoked an enraged Stalin to assign greater priority to the programme. By October 1946, construction had started on Russia's equivalent of 'Windscale' in Chelyabinsk province near the town of Kyshtym, in the southern Urals, conveniently adjacent to a number of hard labour camps. Russian scientists were elevated in status. As one physicist remarked: 'Scientists suddenly became the privileged elite of the country, their living standards ... were raised much higher than the pre-war level.'[35] The arms race, so feared and predicted by the Manhattan Project scientists, was already under way.

* * *

To Attlee's government, proof of the 1949 Russian test came as a double blow: Britain was still at least two years away from testing its own weapon; there were now two superpowers and Britain was not one of them. Post-war Ireland was also coming to terms with its own diminishing relevance in the wider scheme of things. If wartime neutrality had been worn as a necessary badge of independence, payback came in the shape of post-war isolation.

The war consolidated Northern Ireland's place within the United Kingdom, underlining the growing permanence of Irish partition. Anglo-Irish relations quickly settled down into their familiar pattern of tolerance peppered by suspicion, befitting old neighbours who, if not entirely disposed to the warm embrace of mutual understanding, at least knew one another well. A final parting of the political ways came in November 1948 with Ireland's declaration of a republic and formal withdrawal from the Commonwealth. The declaration excited little controversy in Britain beyond predictable anti-Irish hysteria in some of the Tory tabloid press. Ireland's Fine Gael Taoiseach, John A. Costello, who at the head of the country's first inter-party coalition government had replaced De Valera in power earlier that year. hailed the declaration of the republic as 'the end forever, in a simple, clear and unequivocal way [to] this country's long and tragic association with the institution of the British Crown'.

'I believe that as a result of this measure, our relationship with [Great Britain] will be far closer and far better, and will be put upon a better and firmer foundation than it ever has been before,' Costello declaimed.[36]

Immediate post-war relations with the US lay on a different trajectory. The Americans disapproved of Ireland's wartime neutrality. Their simplistic analysis extended to hypocritically looking down their noses at Ireland's failure to come to the aid of Britain in the years it stood alone in the fight against fascism. The damage was exacerbated by Ireland's refusal, again citing partition as the impediment, to join the newly formed North Atlantic Treaty Organisation in 1949.

'The unnatural division of our country is a violation of Ireland's national sovereignty and of the elementary democratic right of self-determination,' was the standard position of the Irish Minister for External Affairs, and former IRA man, Seán McBride.[37] American patience was further tested by diplomatic wrangling over the terms of the Marshall Fund package for Ireland.[38]

The new age of atomic warfare was interpreted by some to mark the end of Ireland's significance in any strategic defence of the Atlantic's western approaches. As the 1948 Berlin airlift sparked off the Cold War and, one by one, the Catholic countries of Eastern Europe succumbed to Soviet domination, the only positive note in official America's perceptions of Ireland was that Irish paranoia about Soviet communism, evident since the 1930s, ascended to new levels. In Catholic Ireland, as throughout Western Europe and the US, Russia was now identified as the enemy.

In the Dáil, members questioned the need for a defence force of 12,000 men and annual expenditure of £4.5m on armed forces in a world dominated by superpowers. 'If the time should come when we would have to fight against atomic bombs, of what use would our little Army, with its rifles, bayonets, trench mortars, howitzers, tanks and lorries be then?' one deputy enquired.[39] James Dillon, the only national political figure to have consistently opposed Irish wartime neutrality, voiced a popular fatalism and the impotence felt by a small country in the atomic age: 'In any future war if we are allied to Great Britain and the United States of America we will come out of the war virtually unscathed,' he said. 'If we are not, we will be blown clean out of the ocean. That is the plain fact. Of course, nobody will admit it. It would be political madness, in this country, to admit that.'[40]

On 15 November 1951, the Waltons entertained Ernest's friend and colleague Professor Wesley Cocker and his wife at their Rathmines home. As the couple bade farewell, Cocker noticed two spent flash bulbs on the Waltons' hall table. He enquired if Ernest was developing an interest in photography. Not at all, Walton replied. The flash bulbs had been left there by a photographer. His nomination for the Nobel Prize in Physics was announced the following day. Walton and Cockcroft received their award in Stockholm on 7 December 1951 for a discovery, the citation read, 'that has profoundly influenced the whole subsequent course of nuclear physics'.

7

Proliferation

'I know not with what weapons World War III will be fought, but World War IV will be fought with sticks and stones.'

Albert Einstein

At a dinner with General Groves at Los Alamos just before the end of the war, the leading Manhattan Project scientists, Oppenheimer, Bohr, Fermi, Chadwick and Lawrence among them, discussed the shape of things to come. Groves expressed concern about the United States' post-war military strength. The scientists speculated about the possibilities of civil nuclear power. Enrico Fermi, Oppenheimer later wrote, mused thoughtfully: 'I think it would be nice if we could find a cure for the common cold.'[1]

Alarmed by the dual prospect of nuclear weapons proliferation and a potentially disastrous post-war arms race, Niels Bohr, in 1944, had asked to see Churchill and Roosevelt to propose a system of international control involving the US, Russia and Britain. Sharing the secrets of the Manhattan Project with the Russians was central to Bohr's plan, which he had discussed with several of his scientific colleagues, including Joseph Rotblat at Los Alamos. Lord Cherwell, present at the meeting with Churchill in April 1944, was embarrassed by Churchill's peremptory dismissal of Bohr. 'I did not like the man when you showed him to me with his hair all over his head,' Churchill later wrote petulantly to Cherwell.[2]

Bohr felt he got along better at a one and a half hour session with Roosevelt. But when Churchill and Roosevelt next met the President agreed with Churchill's suggestion that Bohr be investigated as a Russian spy.[3] At the end of the war, Bohr promptly returned to his Institute in Copenhagen.

Joseph Rotblat was presented with a personnel file alleging espionage activities when he sought to resign from the Manhattan Project in late 1944. Confirmation from Chadwick that army intelligence showed no evidence of a German atomic bomb had finally made up Rotblat's mind. Chadwick advised him to cite the need to try to find his lost wife in Poland as the reason for his resignation rather than principled opposition to the bomb. General Groves agreed he could leave immediately and on Christmas Eve 1944, Rotblat sailed for Europe from New York. A box containing all his scientific papers, which Chadwick had personally loaded onto the ship, disappeared during the course of the voyage, apparently spirited away by American agents.[4]

After the fall of Germany, Leo Szilard sought meetings with high-ranking members of the US administration and circulated petitions among his fellow scientists pleading for the bomb to be scrapped and the existence of the Manhattan Project maintained as a highly classified secret. Edward Teller, later the father of the American hydrogen bomb, likened Szilard's efforts to asking his colleagues to tie a string around the toe of a genie that was already out of the bottle. Szilard's campaign was overtaken by the bombing of Japan. Szilard abandoned physics to concentrate on a campaign to stop the arms race.

Most of the atomic scientists advocated sharing bomb technology and extending international responsibility for control of atomic weapons to the Russians as the only means of avoiding an otherwise inevitable arms race. There were others who felt strongly that disclosure was the least due to a wartime ally who had sacrificed tens of millions of lives in the fight to defeat Hitler. Although the long arm of Stalin's spymaster, the terrifying, psychopathic Beria, stretched deep within the Manhattan Project and Los Alamos and there was no shortage of spies, the majority of the scientists, especially Bohr and Oppenheimer, whose own career would be broken within a few years on the back of spurious evidence of communist sympathies, would have emphatically repudiated betrayal of classified information to the Russians.[5] Trust in the personal integrity of individuals became an early casualty of the new atomic age.

The Nunn May affair had been followed in 1950 by the arrest of Klaus Fuchs. Fuchs, whose family in Germany had been decimated by the Nazis, was an active communist long before he fled to Britain in 1933. His career as a spy began in 1941 and he pursued his double life throughout his Los Alamos years, where he worked on the gaseous diffusion enrichment system for uranium 235. As they observed the technology options favoured by Groves and the military, several of the leading scientists in the gaseous diffusion project had no doubt but that the US was intent on developing a post-war nuclear weapons capability, though it's unclear if Fuchs shared this awareness.

Fuchs coped with his dual persona of scientist and spy, as did the notorious Foreign Office official, Donald McLean, by indulging in regular bouts of heavy drinking.[6] By the time of his arrest in 1950, Fuchs was a naturalised British citizen, working at Harwell on reactor design for the British atomic programme. FBI interrogation revealed his contact, Harry Gold, and the Greenglass–Rosenberg spy ring at Los Alamos. Julius and Ethel Rosenberg were executed in 1953 as a 'deterrent' to other American communists who might contemplate betraying secrets to the Russians.[7] In London, Fuchs was sentenced to fifteen years' imprisonment on 1 March 1950. On his release eight years later, he fled to East Germany.

Notwithstanding spy scandals and defections[8] that soured international relations between East and West as well as exacerbating Anglo-American tensions, not the scientists, nor the goodwill of individual statesmen nor the efforts of the UN Commission established on foot of the Washington Declaration, all put together, could have successfully fashioned a post-war international regime for the control of atomic weapons. The essential factor – a willingness to sacrifice what was perceived as the ultimate guarantee of national security, the bomb, and pool all knowledge related to it to achieve shared international goals – was inconceivable in the immediate aftermath of the war.[9]

The ideological fissure between communism and democratic capitalism, conveniently submerged for the duration of the wartime alliance, had never lain far beneath the surface of relations between the Allies. As the Allied forces had converged on Berlin from East and West in the early months of 1945, Churchill telegraphed his wife Clementine, who was in Russia working with the Red Cross. 'I need scarcely tell you', he wrote, 'that beneath these triumphs lie poisonous politics and deadly international rivalries.'[10]

In post-war America, there was a tendency to view the atomic bomb as God's sacred trust to the American people and frame foreign policy accordingly. In Britain, Attlee's initial espousal of an international control regime, while simultaneously and covertly pursuing an independent atomic deterrent project, was not so much a paradox as an effort to square idealism with the national interest.

The primary purpose of the 1946 Acheson–Lilienthal report, drawn up by a team of consultants that included Oppenheimer, was to put flesh on the bones of the United States government commitment 'to prevent the use of atomic energy for destructive purposes and promote the use of it for the benefit of society'.[11]

The report recommended that all fissile material in the world should be owned, controlled and distributed by a single International Atomic Agency. President Truman appointed the elderly American statesman, Bernard Baruch, to present this US position to a United Nations struggling to agree an international regime to prevent a 'free for all' nuclear arms race. Baruch modified the report to include penalties for breach of any international agreement. In the UN vote on the Baruch proposal later that year, the Soviet Union abstained.

A new international template for foreign relations within which a weapon with the capacity to wipe human civilisation from the face of the earth could be effectively controlled was desirable; but national security remained the overriding imperative. By the end of 1947, the prospect of any international controls regime was a dead letter.

* * *

Attlee and Bevin believed the best hope for post-war Britain and Western Europe lay in a multi-faceted alliance with the US. Without US financial aid and military protection, Bevin, in particular, feared Stalinist hegemony might advance across collapsing economies in western Europe without a shot ever being fired. Like Churchill and other emerging European statesmen of this period, Bevin nurtured ideas of a future European Union. Bevin's concept envisaged a union of European states acting as a counterbalance to the two superpowers. In a pre-war article he had written in 1938 for a trade union journal, Bevin had set out his vision of a loose political and trade alliance among the former colonial powers of Europe and their satellites.[12] The Marshall Plan, which established the Organisation for European Economic Co-operation (OEEC) to draw up a European Recovery Plan and implement the Marshall Aid Fund, represented a means to its partial fulfilment.[13]

To the Americans the Marshall Plan represented something different: an opportunity to extend and deepen American influence, particularly through the creation of a free trade area with a reinvigorated Europe, without the necessity of maintaining a massive defensive military presence on European soil against an encroaching Soviet threat. On the military side, atomic power, dismissed as nonsense only a few years earlier by the chiefs of staff, had by now advanced to the front line of US defence policy.

The Americans used the Marshall Plan as leverage to gain access to Britain's stockpile of Belgian uranium ore and to finally get rid of the much resented and, to most of the US establishment, unacceptable remnants of the British veto on the use of atomic weapons.[14] Another round of secret talks leading to yet another secret agreement on atomic co-operation had been initiated by Britain in December 1947 – so secret that Lord Portal was not allowed to attend the talks in Washington in case his presence aroused comment.[15] The outcome a month later was the Modus Vivendi, aptly described by the then Foreign Office official, and spy, at the atomic energy desk in Washington, Donald McLean, as 'an agreement among people who disagree'.[16]

The Modus Vivendi promised technical information to Britain on the development of civil nuclear power. In return, the US was ceded one-third of Britain's uranium stockpile and all new Belgian Congo ore mined over the next two years. Britain even sacrificed any right of consultation by the US on use of the atomic bomb. Prohibition on providing any atomic development co-operation to Britain's allies in Europe or the Commonwealth was maintained.[17]

Of dubious legality, the Modus Vivendi was never reported to the UN, Congress or Parliament. Nor was it honoured, as far as the British were concerned. The US failed to hand over samples of fissile material and no open-ended visits to the Hanford reactors were arranged as Cockcroft and the Harwell scientists had eagerly anticipated. The Modus Vivendi was curiously silent on atomic weapons development.

Within a few months the Americans were feigning outrage at Britain's admission to making materials for an atomic bomb. Concern was privately expressed in Washington circles about the influence of the left wing of the ruling Labour Party, bolstering the rationale for further restricting technical co-operation. Despite a general recognition of its improbability, the Americans suggested that another European land war, this time with the Soviet Union, might result in the capture of British bomb facilities, a concern reflected in Truman's comment to State Department officials shortly after his re-election to the presidency in 1948. 'We have got to protect our information,' he said, 'and must certainly try to see that the British do not have information with which to build atomic weapons in England because they might be captured.'[18]

The shock of the Soviet's 'Little Joe', the seemingly insatiable appetite of the US bomb production schedule for fresh supplies of uranium and plutonium and the information about Britain's bomb programme, however, rapidly prompted a rethink. With high hopes on both sides and upbeat background briefings from the US administration to the Washington media, talks began in December 1949. This time, the US promised to bring any final agreement to Congress for ratification and notify it to the UN. Fuchs' arrest in 1950, the outbreak of the Korean War coupled with the increasingly pernicious political climate of McCarthyite America, and the US government decision to develop a hydrogen bomb scuppered any prospects of success.

Crucial elements of the proposed deal survived the breakdown of the talks in mid-1951. The Americans were prepared, albeit reluctantly, to accept a British bomb programme as a trade-off for supplies of excess British plutonium. At America's behest, construction was halted on a third pile at Windscale, the foundations of which had already been laid, to conserve international supplies of raw uranium. In return, the US offered American-manufactured bomb components and a stockpile of bombs for Britain. The Americans insisted on continuing the ban on co-operation with third parties, despite British protests that the policy was increasingly alienating its European NATO allies.

British officials accused their American counterparts of a 'dog in the manger' attitude to the exploitation of atomic energy. With masterly diplomatic understatement, the Foreign Office official Roger Makins, later Britain's ambassador to the US from 1953 to 1957, had complained of the 'ill-defined and almost sub-conscious feeling that atomic energy is, and should remain, an American monopoly, both for military and industrial purposes'.[19] The political reality was that, as in Britain, US policy was driven by perceptions of national interest, not the tug of international altruism. The well-developed system of democratic checks and balances through Congress and committees also domestically limited the scope of the executive arm of government, including the President, to make any independent foreign policy commitments.

Attlee's belief that he had secured a personal commitment from President Truman not to use atomic bombs in the Korean War illustrated checks and balances in practice. Dean Acheson, the US Secretary of State, was present at the high-profile discussions in 1950 on use of the bomb in the Far East war, as well as the use of atomic bombs, already accumulating at US bases in the United Kingdom, in any future European war. Acheson claimed he repeatedly cautioned the President not to make any commitments that would limit his authority to act in the US national interest and, in any case, Congress would not permit him to do so.[20] The official communiqué from the meeting was restricted to expressing the President's 'hope' that world conditions would never call for use of the atomic bomb. When the Foreign Office later sought to have their notes of the meeting, including a more explicit undertaking to consult with Britain, accepted as the official record, they were repulsed by the US State Department.[21]

* * *

For a fleeting moment in the immediate post-war era, several of the larger European states, such as Sweden and Italy, considered a military nuclear capacity, and then just as quickly rejected it. Most European or Commonwealth states had no interest in developing atomic weapons, though all shared a keen interest in harnessing nuclear technology for industrial and medical purposes. Civil nuclear power was actively researched in Britain and in the US from 1950 onwards, but by most calculations, the successful development of electricity-generating nuclear reactors was reckoned as somewhere off in the middle distance; at least five, if not ten, years away.

Due to their participation in the wartime project, the Canadians already had a head start on the development of heavy water reactor technology. Post-war Canada had no use for atomic weapons but risked being left with an expensive nuclear white elephant at Chalk River as a hangover from the Manhattan Project. The Canadians transformed the Chalk River site into a reactor development facility for its own civil nuclear programme, but also continued to supply plutonium to the United States – over 250 kgs – up to 1985.

Elsewhere, Britain operated a two-tier policy of limited co-operation, with the 'White Dominions' of South Africa, Rhodesia, Australia and New Zealand occupying the top tier. In the early months of 1947, a joint atomic research programme involving the location of various parts of the atomic project amongst its White Dominions had been considered, but the plan was overtaken by the Modus Vivendi.

South Africa and Australia had the not inconsiderable advantage of substantial natural uranium deposits. The Canadians, with the US and Britain, were part of the Combined Development Agency (CDA), first established under the Quebec Agreement to control uranium supplies. At the end of the war, the CDA controlled 97 per cent of the world's uranium output. In due course CDA contracts guaranteed a measure of British and American technical support for South African and Australian atomic energy projects. Previous colonial ties allied with substantial natural deposits of uranium ensured preference for some countries' nuclear ambitions over others.

Without prior US approval, no British co-operation with non-White Commonwealth states, such as India, or any of the countries of Western Europe, was possible under the terms of the Modus Vivendi. Such approval was only rarely and reluctantly given as the Belgians were to find out. The CDA had signed an agreement in September 1944 with the Belgian government for the entire output of the Shinkolobwe mines in the Congo up to 1956.[22] Under this agreement, as of right, the Belgian government was entitled to a stake in post-war atomic industrial technology, but it took years of protracted negotiations to extract any concessions from the US, who took the view that Belgium should satisfy itself with the profits from raw uranium sales. Most of the information and assistance eventually made available to the Belgians was confined to already declassified information.

Britain, initially poised to become the leader of atomic energy development in Europe, was hamstrung by its 'special relationship' with the United States. The special relationship obliged the UK to turn down a request from France in 1947 to test purified graphite. Similarly, the US refused consent to a request from Norway to Britain in 1948 to barter five tonnes of heavy water for uranium ore.

By 1950, France had emerged as the alternative to Britain for collaborative European research projects. The French, whose programme was always anathema to the Americans because of suspected communist infiltration, independently developed their own heavy water reactor and were engaged in talks with Norway, Sweden, Switzerland and Italy on atomic energy development. A year later, India, snubbed by Britain, concluded its own agreement on technical training facilities with France. In 1960, France became the world's fourth nuclear power, following a successful weapons test in the Sahara.

Britain reached agreement in 1951 with Norway and Holland on the purification of uranium ore for a joint project of which the US was informed, but to which it did not object, only after a year of long-drawn-out negotiations.

Churchill's return to power in 1951 placed Lord Cherwell in charge of the atomic energy project. Cherwell tended to take a more sceptical view of American co-operation than his Labour predecessors or his own cabinet colleagues. When further talks with the US the following year failed to relax restrictions on technical co-operation, Lord Cherwell and his officials adopted their own, less rigid, interpretation of the Modus Vivendi rules and exercised their discretion as to when and where they would seek US consent.

When, in 1952, Belgium requested British expertise to assist in the construction of a small experimental graphite reactor, Lord Cherwell took the proposal to cabinet. Cherwell argued that the US should be informed, but consent for British co-operation with the Belgians was not required. The cabinet agreed with Cherwell. Later that year, Norway proposed the sale of twenty-five tonnes of heavy water, needed at Harwell for reactor research, in part exchange for ten tonnes of uranium metal for its own research reactor. US officials baulked at the deal. The British convinced the Americans that another US 'no' would only drive the Norwegians further into the arms of the French.[23]

Evidence has since become available that from the mid-1950s onwards, Britain regularly exchanged materials with several countries, including France, India and Israel, for use in bomb development without any notification to the US. No use had been found for the twenty-five tonnes of Norwegian heavy water at Harwell. In 1958, twenty tonnes were surreptitiously sold on to Israel by a cash-strapped UK Atomic Energy Authority through a Norwegian company. The heavy water was destined for the French-designed nuclear reactor built by Israel at Dimona in the Negev desert, to make plutonium for its clandestine weapons programme.[24]

* * *

It may seem a rather obvious point that mere possession of all the data or potential nuclear weapons' materials didn't amount to a decision to build bombs. Hans Bethe had remarked in 1945 that there were no big secrets left concerning the atomic bomb, only a lot of little ones. The little secrets could be easily winkled out by competent engineers and scientists given the already universal knowledge of nuclear physics.[25] Any gaps remaining in respect of materials or engineering technology could be filled in with the assistance, usually covert, of like-minded officials or political administrations in friendly states.

The decision to pursue an atomic bomb programme as opposed to limiting nuclear research to industrial and medical applications was then, as it remains today, an entirely political choice. Technology conferred the choice on states of weapons or industrial applications or both.

Public perception on the morality of atomic weapons – drastically different in the US to general European popular opinion – constrained military proliferation in most of Western Europe. The emergence of NATO and subsequent agreements on a defence strategy for the NATO block involving the deployment of US nuclear weapons throughout European bases – the nuclear umbrella – obviated any perceived need for independent deterrent programmes.

Conversely, Europe's dearth of natural energy resources, at a time before free-flowing cheap Middle Eastern oil and gas was even conceived of, made the peaceful development of nuclear energy appear essential as well as attractive. Perceptions of external or regional threats or the need to make a nuclear chest-thumping display in the hopes of being taken seriously as a player in the newly emerging international order motivated the decisions of others.

The atomic genie had long since escaped the bottle and eerily spread its embrace around the globe when Dwight E. Eisenhower, the newly elected President of the United States, made his 'Atoms for Peace' speech to the United Nations on 8 December, 1953. Dag Hammarskjold, Secretary-General of the United Nations, extended an invitation to Eisenhower to address the UN General Assembly whilst the President was engaged at the Bermuda summit with Britain and France in October 1953.

Churchill, obsessed by a personal dread that hydrogen bombs would annihilate human civilisation, was campaigning for a summit with post-Stalinist Russia that he hoped would set the world on the path to international nuclear disarmament. Eisenhower's response, noted by Churchill's private secretary Sir John Colville and leaked to the press by the French, was hardly encouraging. 'Russia was a woman of the streets and whether her dress was new, or just the same old one patched, it was certainly the same whore underneath,' the President said. America would drive Russia off her 'beat' into the back streets.[26]

A month later the President was offering an international summit to Stalin's successors at the Kremlin. His UN address proposed the establishment of the International Atomic Energy Agency (IAEA) to act as a bank for the exchange of nuclear materials and technology and bi-lateral agreements with third countries to engage in the peaceful development of atomic energy.[27]

* * *

At an OEEC meeting in Paris in February 1956, the European states were urged to accept Eisenhower's offer of 20,000 kg of uranium.

Britain's Chancellor of the Exchequer, Harold Macmillan, who chaired the meeting, spoke of atomic energy as 'a new opportunity of European co-operation ... an opportunity where Europe can get together ... and strike the imagination of our people'.[28] The EURATOM group of six states, later founder members of the European Economic Community, while aligning themselves with the OEEC plan reserved the right to move ahead independently of it.

Liam Cosgrave, Minister for External Affairs, attended the Paris meeting on behalf of the Irish government. The following day, 1 March 1956, the *Irish Times* reported: 'Mr. Cosgrave's reference to "co-operative enterprises" was interpreted in Dublin last night as an indication that the government has decided to accept the offer of the United States government to help Ireland acquire an experimental nuclear reactor.'[29]

John A. Costello had again been asked to lead the second inter-party coalition government in Ireland, formed after the general election of 1954. After more than thirty years of independence, if the state could be said to be efficient in one thing it was in the fostering of elites: among the farmers and professional classes from which they themselves were mainly drawn, as well as social cliques of public servants, small shopkeepers, a shamelessly irredentist trade union movement and a lower middle class; all fiercely defensive of their heritage and privileges, scant as they were.

An Irish micro-universe in which cultural repression, religion and moral superiority masqueraded as the fulfilment of the nationalist promise of self-reliance had been created. The Ireland of the 1950s, in short, gave a fair impression of a society bound in self-sustaining stasis, replete with smug satisfaction in the meanness of its achievements. Richard Mulcahy, Minister for Education in the first inter-party government in 1950, could claim without creating a ripple of public unease that: 'In the world of today advance of knowledge has led to endless destruction and misery.' 'The foundation and crown of youth's entire training is religion,' Mulcahy opined.[30]

Such views may have reflected an elitist consensus of the time but a few liberal-minded and forward-thinking politicians and civil servants were increasingly waking up to the fact that the country exhibited all the characteristics, with the exception of its moribund political stability, of an otherwise failed state.[31] However well-prepared their souls might be for the life hereafter, the poorly skilled and half-educated masses forced to emigrate in their thousands were ill equipped to meet the challenges of the modern world.

Despite the disappointments of actual performance, the state retained aspirations to a place for Ireland in the modern world. Writing in confidence to Professor Ernest Walton on 29 February 1956, Taoiseach John A Costello announced his government's decision to establish an 'Atomic Energy Committee' (AEC) and asked Walton to accept a year's appointment as a member of the committee. The purpose of the AEC was to advise the government on 'the various peaceful uses that can be made of atomic energy in Ireland'.

The nine-member committee was given a remit to assess Ireland's economic requirement for atomic energy, the type and location of a research reactor including 'the provision of safeguards against radioactivity',[32] and the training and education needs of Irish engineers and physicists. Even before the OEEC Paris meeting, the government had already decided in principle to accept a US offer under the Atoms for Peace programme of 30 kg of enriched uranium for research purposes.

Early in March, on the same day the *Irish Times* reported a meeting of the Engineers' Association in Merrion Row supporting the development of atomic energy in Ireland, Costello visited the US nuclear research reactor at Brookhaven, New York.[33] Cosgrave met with a delegation from the Engineers' Association. On St Patrick's Day, 17 March 1956, Ireland's ambassador to Washington, John Hearne, initialled the bilateral agreement at the US State Department in Washington.

At its inaugural meeting on 15 April 1956, the Taoiseach described the AEC as 'the first step in introducing Ireland to the atomic age'. Nuclear energy, he said 'must rank as one of the greatest technical achievements of all time and one of the most opportune'.

'This Committee may therefore consider how soon the demand for power in industry, agriculture and the home will outstrip our present resources and the role which nuclear energy could play in meeting the expanding demand,' Costello said.[34]

As it turned out, the AEC experienced great difficulty in reaching agreement on the need in Ireland even for a research reactor. Ernest Walton, for one, was highly dubious about the entire project.

In a letter on 26 April 1958 to John Conroy, the Irish Transport and General Workers Union representative on the committee, Walton wrote:

> How much of our national income should be invested in science for the sake of the future is a matter of discussion. I am convinced that we are not spending nearly enough at present and

that part of our present plight is due to our failure in this respect in the past. However, I would not put a research reactor anywhere near the top of my list of scientific priorities.

In his private papers, Walton estimated the likely cost of a nuclear research reactor at £330,000.[35] Outlining his reservations to the committee, Walton stressed that the case for the 'immediate purchase of a research reactor would be strengthened immensely if it could be shown that it would be likely to give results of importance to large existing industries or which might lead to the establishment of new ones'.[36]

In Walton's view, it couldn't. By the time the AEC reported to the government in May 1958, with a majority view in favour of the reactor proposal, De Valera's Fianna Fáil was once again in power. A year later, De Valera announced the government's decision not to acquire a research reactor. Responding to Dáil questions from Liam Cosgrave, De Valera said: 'I think the best way to put it is that it would not be justifiable at the moment to spend that sum of money. There are changes taking place and the immediate need for the reactor is not there.'

In an indication of the parlous state of the Irish economy at that time, De Valera also announced restrictions on nuclear medicine. The government's view, he stated, was that it 'would be undesirable to establish units for the therapeutic use of radioactive isotopes in general hospitals; and facilities for the use of radioactive isotopes for diagnostic purposes should not be extended to hospitals outside Dublin'.[37]

For the present at least, an impoverished Ireland would forgo any benefits of nuclear technology.

8

Heat and Light

'It is not too much to expect that our children will enjoy in their homes electrical energy too cheap to meter, will know of great periodic regional famines in the world only as matters of history, will travel effortlessly over the seas and under them and through the air with a minimum of danger and at great speeds, and will experience a lifespan far longer than ours ... This is the forecast for an age of peace.'

Lewis Strauss, USAEC Chairman, 1954

The tenant of Calder Hall surrendered the lease on his farm to the Ministry of Supply early in 1953. Later that summer, construction began on the first of four dual-purpose gas-cooled reactors. Calder Hall, the first commercial-scale nuclear power plant in the Western world, was officially opened by the Queen on 17 October 1956, much to the delight of the nuclear establishment and government in Britain, who could at last lay claim to an international landmark in atomic energy development, and chagrin in the US at being pipped at the post by their erstwhile 'junior partner'.[1]

The gas-cooled Calder Hall reactor was hardly the optimum design for the development of large-scale commercial nuclear power. Actively engaged in nuclear power research from the late 1940s, the preferred option among Harwell and Risley scientists and engineers, like their counterparts around the world, was for a fast-breeder reactor, using plutonium as fuel.[2] The problem was that successful fast-breeder development was reckoned to be at least twenty to thirty years away. Lack of resources limited other options. Since an experimental reactor cost between £1m and £2m to build, development costs restricted the number of designs that could be actively researched.[3]

As Britain's chief of production, Christopher Hinton had considered the feasibility of a joint Anglo-Canadian effort to develop heavy water-moderated reactors. Britain had no indigenous heavy water supply, and a manufacturing plant for heavy water, the scientists concluded, would probably pose as many environmental hazards, especially the risk of explosion, as a Windscale pile. The US practically held a monopoly on the production of enriched uranium, essential to light water reactor designs. This meant that unless Britain could produce an adequate supply of its own enriched uranium, none was likely to become available in the medium term.

The Calder Hall design – graphite-moderated reactor within a containment vessel, fuelled by natural uranium and cooled by pressurised carbon dioxide – built on existing technology with which the development team were already familiar from the Windscale piles. Raw materials were readily available in Britain. While Lord Portal had ruled out Hinton's suggestion in 1947 that one of the original Windscale reactors should be gas-cooled, he had encouraged further research and development. The prototype, Pippa, designed and built at Harwell, was originally conceived as a nuclear power plant that produced plutonium as a by-product.[4]

The scientists and engineers on the development team had no illusions about the relative costs of nuclear power or its capacity to compete with or replace coal as Britain's primary energy source. From the beginning, the high capital costs of a Pippa-type power plant were recognised.[5]

'We do not expect to produce a cheaper source of power than that derived from coal – it is likely, in fact, to be somewhat more expensive,' John Cockcroft had remarked in 1951. 'What we are aiming at is to increase the total power available.'[6]

Natural uranium might compete with coal as an energy source in 1950s prices, but the long-term economics of nuclear power depended on a range of variables, including the relative availability or scarcity of coal and oil supplies, inflation and international currency values. B.L. Goodlet, the Risley engineer who designed the Pippa reactor, dismissed the notion of nuclear power ever providing cheap electricity. Power generation from the small Pippa reactor, he estimated, would cost four to five times the equivalent of a coal-fired station.[7]

Nor was there any euphoria about nuclear emerging as a dominant energy source. Cockcroft was sceptical of any large-scale development of nuclear energy for at least twenty years. Hinton described a 1949 paper by an economist in the Ministry of Fuel and Power, suggesting that nuclear power could replace most existing power supplies in Britain by 1965–70, as akin to 'an H.G. Wells novel'.[8]

The real economic advantage of Pippa lay in not having to build any more Windscale piles. The heat produced in the Windscale reactors simply went up the chimney. Electricity bought from the national grid to air-cool the piles cost an average £680,000 per annum.[9] In contrast the costs of irradiating nuclear fuel in a Pippa could be offset by the value of the military plutonium subsequently extracted. The electricity then produced, at least in theory, should cost nothing. Sold on to the national grid, it represented pure profit.[10]

Calder Hall also offered a short-term solution to a more immediate problem – the shortage of plutonium for the bomb programme. The capacity of the Windscale piles to produce plutonium had been overestimated in the original physics calculations. In August 1952, as the Cold War hardened, the chiefs of staff demanded the doubling of plutonium production within the next three years, to meet their target of a stockpile of 200 atomic bombs by 1956. To Lord Cherwell's horror, if only because of the capital cost involved, the chiefs of staff suggested building a new plutonium production plant in the Commonwealth. Alternatively, the plan for a third Windscale Pile at Sellafield, abandoned following the Anglo-American talks on co-operation in 1949, could be revived or an entirely new reactor design brought forward. Lord Cherwell met with the chiefs of staff. If they were prepared to be patient and extend the deadline for doubling plutonium production from three to four years, he would solve their problem.

Cherwell's solution involved tweaking the Pippa design so that its primary purpose became to produce plutonium with electricity as a by-product, rather than the other way round. The Windscale piles had been designed as low-temperature reactors to maximise the production of plutonium 239, or weapons-grade plutonium, in irradiated natural uranium. A power reactor, however, running at a much higher temperature to produce electricity efficiently, would produce large quantities of plutonium 240, commonly known as civil-grade plutonium, that was limited in its usefulness for military purposes. Adapting the Pippa design to produce more plutonium 239 involved changes to the dual-purpose heat converter and the installation of larger boilers and turbines.

Although he was not in a position to be precise about the technology, Cherwell put his outline proposal to his fellow ministers on the Defence Committee in December 1952, two months after Britain's first atomic bomb test at Monte Bello. Churchill, preoccupied with plans to meet the newly elected President Eisenhower and reinvigorate talks on international nuclear disarmament, abruptly refused to countenance any increase of plutonium production in Britain.

Cherwell could scarcely contain his anger. 'I cannot believe you would contemplate adopting such a disastrous line which might well in the long run spell national suicide,' he wrote in a memo to the Prime Minister.

Cherwell argued that progress on the development of nuclear power was dependent on the production of materials for atomic bombs. Against an overall defence budget of £1.5bn a year, the extra cost of the nuclear programme at about £6m a year for the next four years was hardly significant.

'Unless we are to sacrifice all hope of holding our own in this vital field of exploitation of nuclear energy we shall have to spend almost the same amount of money whether we make atomic bombs or not,' he stated.[11]

The Calder Hall project was approved in February 1953, as was Cherwell's scheme for a UK Atomic Energy Authority, established by statute in 1954, to take over control of all atomic energy development from the Ministry of Supply. Shortly before his resignation as Prime Minister, Churchill, in what has been described as his 'last will and testament' speech to the House of Commons on the defence budget on 1 March 1955, announced the decision to build a hydrogen bomb.

This decision was inspired as much by hopes of increasing leverage on the US for renewed atomic co-operation as a signal of deterrence to the Soviet Union.[12]

'You can imagine what my thoughts are about London. I am told that several million people would certainly be obliterated by four of five of the latest H bombs. In a few more years these could be delivered by rocket without even hazarding the life of a pilot,' Churchill had written to Eisenhower on 4 March 1954.

'Perhaps we have now reached, or are reaching, the moment when both sides know enough to outline the doom-laden facts to each other,' he said.

The Strath report, commissioned by Churchill's government a year previously, had laid out the devastating consequences of a hydrogen bomb attack on Britain. Ten bombs would kill twelve million people in the initial blast, with four million others seriously injured, or about one-third of the population. One bomb striking London might account for four million of the casualties. A copy of the report had been furnished to every member of the cabinet.[13]

By May 1955, with all hopes for a multi-lateral disarmament scheme long since faded and Anglo-American atomic co-operation at an all-time low, the cabinet sanctioned a scheme for construction of four reactors at Calder Hall and a further six at Chapelcross in Scotland to meet military plutonium production requirements.

* * *

Not a whisper of the military rationale for the Calder Hall reactor programme disturbed the fanfare of its sumptuous official opening in October 1956 and the connection, for the first time in the world, of nuclear-generated electricity to Workington, a town fifteen miles north of the Windscale site. Swords into ploughshares dominated the Queen's official opening speech, comments by dignitaries and statesmen, and subsequent coverage in the world's media. The London *Times* spoke of the 'magic' of this development, while the *Daily Telegraph* pronounced: 'Calder Hall has started a new age.'[14]

A report in the *Irish Times* liberally quoted the Queen's address, including its accolade of 'limitless opportunities' and self-congratulation on the harnessing of the atom for peaceful purposes: 'the greatest of our contributions to human welfare'.[15] The Queen's lunch at Calder Hall, the paper reported, was 'the first royal meal to be cooked by atomic electricity'.[16]

British officials at the opening of Calder Hall had pronounced that after 1965, every new power station built would be an atomic power station.

The UK government's 1955 White Paper, *A Programme for Nuclear Power*, proposed commissioning twelve nuclear power plants to supply 1,500 to 2,000 MW of electricity by 1965. In the wake of the Suez crisis and the humiliating débâcle of the joint British and French invasion of Egypt in 1956, followed by a temporary crisis in oil supplies from the Middle East, this target was revised upwards to 6,000 MW. Ten years later, as fears about energy supplies diminished, it was revised downwards to 5,000 MW.

Magnox reactors, so called because the uranium fuel elements were encased within a magnesium oxide cladding, were regarded as a proven design. Scaled up, they offered the safe and reliable production of nuclear power, making a valuable addition to Britain's energy mix and its security of supply. Magnox reactors had a further advantage: they would also produce plutonium as a by-product.

In reply to a parliamentary question on 1 August 1957, the Paymaster General stated that three of the new 900 MW civil magnox reactors then in the process of being designed would be modified to enable them to operate on a military fuel cycle, and to produce between them 1,200–1,400 kg per annum of weapons-grade plutonium from 1963 onwards.[17] Although the military cycle plan was later confined to two reactors on the Hunterston site in Scotland, any dividing lines between the civil and military programmes were blurred, especially since the fuel for both the civil and military cycles was reprocessed in the same plant at Sellafield.

* * *

In the context of Tube Alloys and later the Manhattan Project, secrecy was understandable as a necessary part of security in time of war. In the immediate post-war period, political leaders in Britain and the US could plead proliferation as a reason for temporarily keeping all materials and information relating to the Manhattan Project highly classified. In fact, a basic description of all stages of the Manhattan bomb project, the Smyth Report, albeit in the teeth of British official opposition, had been published as an appendix to the presidential statement that followed the bombing of Hiroshima. Christopher Hinton claimed the Smyth Report proved very useful when he was struggling with the design of the Windscale plants. General Groves had approved the report's publication in a tactical move to appease the project scientists. Otherwise, Groves feared they would take matters into their own hands and publicise details of their work on the project that he preferred should remain unknown.

Many of the scientists recruited to the post-war British project by Cockcroft and Chadwick wanted the strictures to which they had been subject throughout the Manhattan years relaxed and a guarantee of permission to publish research results in scientific journals. In time, and perhaps chastened by the series of spy scandals, they became more habituated to the overall climate of secrecy. Some scientists, and particularly engineers employed at Risley, subsequently claimed to have no understanding that their work was connected to anything other than research into a new system of energy; not that it was directed towards the production of atomic bombs.[18]

The risk of alienating Russia appeared reason enough for keeping the existence and extent of Anglo-American wartime secret agreements under wraps while the prospect of an international control regime was under investigation by the UN, as did the risk of revealing atomic secrets to the Russians that would help them secure the bomb once the international control effort finally faltered. Differences in democratic processes and political culture between Britain and the US – and respective public attitudes to atomic weapons – propelled the Americans in the direction of imposing the most draconian sanctions on individuals who transgressed the directives or even the spirit of the McMahon Act, while at the same time deluging scientific journals, the general media and the public with as much information about atomic energy as they felt they could safely declassify.

From the beginning, the US administration involved private corporations in all industrial aspects of atomic energy development which, in itself, forced a more open discussion of the uses of atomic energy. In Britain, the project was retained under government control partly because there was only a small reservoir of suitable firms available to which projects could be outsourced. Most major British firms, like ICI, declined to become involved other than in a relatively minor sub-contracting role. There was a belief, especially among the scientists and at a high political level, that the involvement of private industry in a national project of such prestige would be inappropriate.[19]

At the 1949 talks, the US suggested that the British could do with a few good PR men. The British delegates felt the Americans could do with dampening down the average American's naiveté about atomic weapons and cavalier enthusiasm for their use, as evidenced in the disturbing level of emerging public support in the US for the alarming concept of 'preventive war'.[20]

Officially, Attlee and Bevin sought to justify their almost irrational insistence on absolute secrecy and a total clampdown on any publicity relating to the British programme to a fear that release of even the most innocuous information might help the Russians or, less openly acknowledged, reveal to the Americans the small-scale, resource-starved limitations of the British programme. Undeniably, there was also a real concern about the impact that public disclosure of Britain's bomb-making, and diversion of scarce resources to that end, might have within their own party and among loyal Labour supporters.

In the early post-war years of 1946 and 1947, Britain endured a disastrous combination of appalling winters and coal shortages, resulting in many otherwise avoidable deaths from hypothermia. The increasingly unpopular rationing of basic foodstuffs remained in place. Labour voters could not be relied upon to share their Prime Minister's and Foreign Secretary's conviction that the national interest made construction of atomic bombs a priority.

* * *

Joseph Rotblat, settled into a position as a teaching Professor at St Bartholomew's Hospital in London, founded the Atomic Scientists Association in 1946, with the aim of informing the public of the peaceful and beneficial uses of nuclear technology as well as alerting them to its dangers. In autumn 1947, the association organised a travelling train exhibition. Ernest Bevin was not impressed by this proposed PR exercise; exhibition photographs were withdrawn. Bevin directed that, in future: 'Nothing must be allowed to happen without both the Prime Minister and myself [being] informed.'[21]

Attlee, almost habitually, gave a negative response to any proposals for media visits to Harwell. Permission to nationally disseminate press information and photographs in June 1950 of the laying of the pipeline to discharge waste from Windscale to the Irish Sea was refused, even though there was no secrecy surrounding the project and the details were widely known locally in Whitehaven.

Throughout Labour's entire period in office, no House of Commons debate on Britain's atomic programme took place and all parliamentary questions were directed to the Prime Minister for answer.

'When an Honourable Member asks the Prime Minister about the atomic bomb, he looks at him as if he had asked about something indecent,' the Scottish Labour MP Emrys Hughes, observed.[22]

The media were muzzled through a combination of official D-notices prohibiting the publication of specific information in the press and the supineness of journalists and editors themselves. The D-notice that accompanied the 1948 atomic bomb announcement in the House of Commons ruled out: disclosure or reference to locations in the UK where development or production of atomic weapons material was in progress; bomb design and materials used; description of bombs and materials storage facilities or the identity of personnel working on atomic weapons. A second D-notice early in 1951 prohibited any media comment on the appointment of Sir William Penney as head of the Aldermaston site.

In August that year a *Daily Telegraph* article mentioned a bomb test in Australia. Malcolm Muggeridge, later a renowned broadcaster, then deputy editor of the *Telegraph*, sent an abject apology to the Press Committee monitoring the implementation of D-notices: 'You may be sure the lapse was unintentional and that all requisite steps will be taken to avoid any repetition of the offence,' he wrote.[23]

If anything, the climate of official secrecy intensified with the return of the Conservatives to power in 1951. Suggestions for a public information campaign by John Cockcroft and others in summer 1952, including publication of a White Paper on the atomic bomb project and separate brochures outlining the functions of Harwell and Risley, were received with little official enthusiasm.

'Frankly, I personally dislike publicity,' Cherwell remarked, although he authorised the production of a booklet on Risley that finally appeared in 1954. Cherwell further conceded that aspects of the D-notice system, suppressing publication of information and ubiquitous 'no comment' responses to obvious facts, sometimes made the government appear 'slightly ridiculous'. But the US was operating in a different environment, he surmised, and also benefited from 'the absence of any political opposition from pacifists or pseudo-Communists'.[24]

Events such as the Monte Bello test in 1952 or the opening of Calder Hall attracted a flurry of publicity that was all froth and no analysis and tended to die away relatively quickly. By and large, the national media in Britain relied on the American press for atomic energy stories, including the revelation, for example, of a secret meeting between Attlee and Churchill during the troubled Anglo-American talks in 1949.

Apart from some occasional articles in the *Manchester Guardian* throughout this period, no meaningful analysis of the British government's nuclear deterrence strategy was attempted in the media. Instead, public perception was left to the mercy of overblown hype and futuristic fantasies about atomic power filtering through from the US media, not least the subsequently infamous claim by the USAEC Chairman, Lewis Strauss, about 'electricity too cheap to meter' delivered at least two years before the US own civil nuclear power programme demonstrated any commercial potential.

* * *

If the net result of official and government refusal to engage with public opinion on atomic energy policy was to induce a welcome temporary mood of public complacency, in the longer term it served only to shore up hostages to fortune, especially in the areas of radiological safety and health.

From 1948, the UK had a basic regulatory framework, which, among other things, prohibited the then still relatively common practice of including radioactive ingredients in cosmetic preparations, such as hair dyes or facial mudpacks. Scientists had succeeded in establishing the International Commission for Radiological Protection (ICRP), which developed guidelines for safe doses of radiation. A system for ensuring radiological protection of workers on the various sites, and especially Windscale, was in place from the early days of the project. Both Cockcroft and Hinton vigorously resisted attempts to reduce the number of health and safety personnel on their sites in the general round of cutbacks on public expenditure pursued by the Conservative government on their return to power in 1951.

Cockcroft and Hinton also shared misgivings about the lack of public information and political debate on Britain's atomic energy policy. They believed the overwhelming secrecy surrounding all aspects of the project must ultimately prove counter-productive. On a practical level, secrecy fostered a lack of joined-up thinking between departments. Up to 1954, the atomic bomb project was pursued outside the mainstream of defence policy, and bombs were produced without any thought given to the types of aircraft necessary to deliver them. There was no public accountability for expenditure on the project, diffused and concealed as it was within general defence appropriations.[25]

The active suppression of any public probing or debate on the rationale for the bomb programme promoted the growth of an organisational culture within the UK Atomic Energy Authority (UKAEA) that rarely questioned its own motives. This strange hybrid within the British system of a publicly owned body with a private sector commercial mandate, a secret military function allied to a civil nuclear programme, the UKAEA creaked under the weight of its internal contradictions.

Operationally, it could never learn from its mistakes because secrecy ensured it was never put under any public pressure to acknowledge any mistakes. Frank exchange of information, even internally, on issues such as the production and storage of a growing volume of high-level radioactive wastes and low-level radioactive discharges to the Irish Sea, both later central to Ireland's dispute with Britain over Sellafield, was also constrained.

* * *

At a meeting with the Minister for Supply and his Permanent Secretary at Risley in April 1950, Christopher Hinton gave an impassioned account of his concerns about a potential catastrophic accident involving the six tanks under construction at Windscale to hold the highly active liquid waste (HAL) derived from reprocessing irradiated uranium to extract and refine plutonium. The tanks of specially welded stainless steel with a biological shield of concrete, lined with more stainless steel, were designed to ensure containment of the highly radiotoxic waste that would be released to the atmosphere if the tanks overheated and exploded or if there was any significant leak of the material. When full, the six tanks would contain five million curies[26] of radioactivity. Apart from the ever-present risk of a serious accident,

Hinton aired the long-term problems he foresaw with the effective management of this waste and creation of an 'historic legacy' that in time would pose huge and expensive problems of storage and disposal.[27]

Rather exceptionally for a politician, the Minister for Works, Richard Stokes, had raised the HAL safety issue with the Ministry of Supply. The ministry's formal response to his queries reflected none of Hinton's concerns. The letter reviewed the safety features of the Windscale tanks and, in what was becoming an honoured tradition of administrative bullying in response to salient questions on the atomic programme, concluded that the only way to ensure total safety was to abandon the bomb project altogether.[28]

From the beginning, it was also a given that Windscale would produce massive volumes of low- and medium-activity liquid wastes, much of it water from the storage ponds in which irradiated uranium was stored prior to processing for plutonium.

'The sea has always been regarded by coastal and seafaring peoples as the ideal place for dumping their waste and this is, of course, a very reasonable and proper attitude,' the UKAEA physicist John Dunster, who later became the UK government's chief advisor on monitoring the environmental impact of both the Windscale and Chernobyl accidents, told a UN conference in Geneva in 1952.

Among the range of experiments conducted to determine the distance to sea at which the pipeline carrying low-activity liquid from the site should be laid,[29] some 35,000 fish were marked to study population behaviour in the Irish Sea. Almost half the marked fish were caught and examined for contamination levels, providing data that showed there would be considerable dilution of radioactivity in sea areas remote from the site. Fluorescent tracers were used to investigate the movement of water in an experiment known as 'Seanuts'. The highly visible tracers used in this experiment raised eyebrows in Whitehaven, creating problems for the local site management. Hugh Gethin Davey, the Works general manager, was embarrassed locally by his lack of any prior knowledge of the experiment or its purpose.

Plutonium and other fission products were released to the Irish Sea to measure their relative concentration in marine organisms and fish, the most obvious and inevitable 'pathway' to human consumption, and to determine an acceptable balance between discharge levels and human dose limits. The experiments showed, for example, that ruthenium concentrated particularly in edible seaweed, which was harvested locally along eighty miles of coastline and sent to Wales to be made into the Welsh delicacy, laver bread.[30]

'The intention has been to discharge fairly substantial amounts of radioactivity as part of an organised and deliberate scientific experiment,' Dunster told the Geneva conference. In words with which anti-nuclear lobbyists would taunt him for the rest of his life, even in obituaries following his death in May 2006, Dunster lauded the lack of administrative controls that made the Irish Sea such an attractive dumping ground for Windscale's radioactive wastes. 'The aims of this experiment would have been defeated if the level of radioactivity discharged had been kept to a minimum,' he said.[31]

Data from the experiments were used to calculate site discharge limits for particular radionuclides to minimise the uptake of radio-activity by consumers of fish and seafood, and in Welsh laver bread. The data also determined the distance for safe marine discharges. A double pipeline was laid two miles out to sea in a ten-day operation that began on 1 June 1950. The first routine discharges of radioactive effluent to the Irish Sea began in February 1952.

Windscale, its management had decided early on, was a chemical works with an additional substantial radiological component, as reflected in the Health Physics and Safety Department established in 1948, two years before the first pile was expected to become operational. This alliance of conventional safety practice with radiological monitoring employed ten physicists and sixty health physics monitors in 1950. The UKAEA purchased and equipped its own vessel, the *Mary Munroe III*, to monitor the marine environment, collecting samples of sand, silt, fish and the seabed for monthly analysis. On shore, air samples and local vegetation were similarly subjected to radiometric analysis from 1950 onwards.

* * *

Visiting Harwell in 1949, Edward Teller suggested that the British scientists had exaggerated the dangers of water-cooled reactors. In his opinion, graphite-moderated reactors, such as those under construction at Sellafield, carried a far greater accident risk because of the presence of so-called 'Wigner energy' in the graphite lattice. Wigner energy is the potential energy that can build up in graphite in reactors running at low temperatures, and that can then be suddenly released. If hot spots of 'Wigner energy' were not discharged at regular intervals, they could, theoretically at least, result in a spontaneous graphite fire. Leo

Szilard had first identified this phenomenon in graphite reactors, and his fellow Hungarian, Eugene Wigner, had given his name to it and to a related problem, the 'Wigner effect' in which stray neutrons eventually caused graphite to buckle and distort, which in turn affected the physical integrity of the pile.

The risk of reactor fire resulting from oxidisation of burst uranium cartridges was already well known to the scientists. Teller believed that one burst cartridge, in combination with stored Wigner energy, would be enough to ignite an entire pile.[32]

Teller's insights were not communicated to the engineers at Risley nor to the operators at the Windscale site. When an unexpected temperature rise was observed in Pile No. 2 on 7 May 1952 its cause was not understood. The problem was dealt with by increasing the flow of cooling air through the pile. Later in September 1952, when Pile No. 1 was shut down for routine maintenance, another unexplained rise in temperature occurred and smoke was seen coming from the reactor. The smoke turned out to be innocuous – oil from the cooling blowers had been carried into the core. Teller's predicted spontaneous release of Wigner energy had occurred in both reactors.[33] An annealing process, which involved heating the reactors from the base up in order to free the trapped energy, became a routine part of general pile maintenance. Unfortunately, the process was never the same and defied rule book definition. Worse, pockets of 'Wigner energy' sometimes remained trapped in the reactor core.

* * *

The best thing that could be said about the Windscale piles was that they showed how not to build a reactor. Expensive to operate, wasteful of scarce raw materials, and inefficient even in their primary purpose of plutonium production,[34] they were also slowly poisoning the surrounding countryside.

Burst cartridges, if they remained undetected, oxidised slowly and released radioactive particles into the atmosphere through the stacks. The filters at the top of the piles, known on the site as 'Cockcroft's follies', were almost entirely ineffective in capturing these small particles.

In 1952, 140 displaced cartridges were detected in Pile No. 2. In July 1955, radioactive 'hot spots' were found in the area up to three miles surrounding the reactors. The July 1955 survey for the first time used on-the-ground probes to detect radioactivity. It showed that the routine air-sampling system for gamma radiation, which depended on collecting samples from a series of given points around the site, was defective: it masked the presence of the 'hot spots' caused by particulate emissions from the stacks. The immediate contamination source was traced to thirteen failed uranium cartridges in Pile No. 2, displaced and lodged in the air ducts at the base of the pile. The survey revealed a significant amount of the contamination was more than two years old.

The UKAEA board met to discuss this incident on 27 September 1955. The meeting was attended by representatives of the Ministries of Agriculture, Housing and Local Government. A public statement was considered.

'The view was expressed that it would be unfortunate if no statement was now made and we were obliged to admit that the Authority had kept this matter quiet and had to have the information extracted from them,' the minutes recorded.

'On the other hand it was pointed out that there seemed to be no biological risk at all, that the public would only be puzzled by a statement that something had happened in which there was no danger and that any statement that was made was likely to lead to public misunderstanding.'

The UKAEA appointed a medical panel to assess the radiological significance of the particle emissions. Two months later the Authority's Board was advised that there was no danger to the public. An engineering expert team who visited Windscale in September 1955 recommended the development of an improved filter system to capture the particles.

But it was no use. Health Physics' survey results in January 1957 indicated three separate incidents of particle emissions in the previous twelve months.[35] Following a visit to Windscale in July 1957, the Medical Research Council (MRC) reported: 'We were informed that the emission of active particles in the cooling air from the pile stacks, which was first discovered in 1955, does in fact continue steadily.'

The MRC proposed a large-scale biological monitoring programme. Levels of strontium 89 and 90 in local milk samples were found to be 'two-thirds of the permitted levels'. The findings raised concerns in the Ministry of Agriculture and were reported in a note to the Prime Minister. The UKAEA was placed on notice that if the contamination continued, there would be no alternative but to close down the piles. The life of the Windscale piles, which Hinton had estimated could be no more than five to ten years in any case, was drawing to a close.

9

The Windscale Fire, 1957

'Most of the radioactivity was blown out to sea.'

UKAEA Press Release on Windscale Fire

A primary goal of Britain's post war foreign policy – an atomic co-operation agreement with the US – remained stubbornly elusive throughout the 1950s. The replacement of the 1946 McMahon Act in 1954 by a new Act provided for agreements on atomic co-operation that ultimately only heightened British frustration. The new atomic energy legislation exacerbated the deep-rooted political conflict within the US Congress as to whether development of civil nuclear power should be federally controlled, as the Democrats believed, or left to private enterprise, as favoured by the Republicans.

Rivalry between the politicians on Capitol Hill, especially the members of the secret Joint Committee on Atomic Energy (JCAE) and the US Atomic Energy Committee (USAEC) over control of nuclear policy, also militated against any effective agreement with third countries. Agreement by the NATO Council in 1954 to incorporate US nuclear weapons into war plans had been followed a year later by Anglo-American agreements on civil and military co-operation on atomic energy. But domestic political tensions between Congress and the White House negated their effect.

By 1957 the Americans appeared sufficiently impressed by British development of an independent nuclear deterrent, the start of the civil nuclear power programme and the limited tests in Australia on development of an H-Bomb to open the door a chink on British aspirations to a transatlantic deal.

Eisenhower was at last persuaded by the British Premier, Macmillan, who had been his secretary for a period during the Second World War, to bring an amendment of the 1954 Act to the floor of Congress that would facilitate the long-sought co-operation agreement between the two countries. High-level diplomatic discussions were in progress on military technology exchange and, as always, the Americans were keen to secure supplies of surplus British plutonium for their own stockpiles. The launch of Sputnik I, the first artificial satellite, by the Russians on 4 October 1957 signalled the age of intercontinental ballistic missiles, adding a spur to US enthusiasm for a new transatlantic arrangement with its closest European ally.

* * *

Ironically, the secrecy that enveloped the British project had worked in its favour among recalcitrant American senators and congressmen. In the US the 1952 Monte Bello test had been widely reported as superior to any United States bomb – which it clearly was not. Even if influential sections within the US nuclear establishment were not fooled, enough confusion was engendered on Capitol Hill about the status of the British programme to promote broader diplomatic objectives.[1]

Chadwick had always counselled leaving the Americans alone and for Britain to get on with the business of developing its own bomb as the best means of ultimately securing the joint co-operation with the US for which British politicians so earnestly strove. Writing just prior to the Monte Bello test, William Penney summed up the general British view: 'The discriminative test for a first class power is whether it has made an atomic bomb,' he said. 'We have either got to pass this test or suffer a serious loss in prestige both inside this country and internationally.'[2]

A first-class power in hot pursuit of the final consummation of its special relationship with the United States had a vested interest in drawing a discreet veil over any deficiencies in the execution of its independent deterrence programme. Most likely for this reason, the Prime Minister had personally ordered that Windscale's pollution of the local Cumbrian environment throughout the 1950s should remain secret.[3] But however inconvenient the timing from a political perspective, no possibility existed of concealing the major accident that occurred in early October 1957.

* * *

Up to the opening of Calder Hall in 1956, the entire British military and industrial programmes depended on the successful operation of the Windscale piles. Plutonium for bombs was the first priority. The piles were also under pressure to produce an increasing array of artificial isotopes for use in medicine, industry and agriculture in Britain and for export to Europe.[4] Hydrogen bombs required substantial quantities of tritium, a hydrogen isotope with an atomic mass of 3.

Polonium 210, used as a triggering device in conventional atomic bombs, was also produced in Windscale in so-called LM cartridges. The cartridges used to manufacture tritium were of a lithium–magnesium alloy, codenamed AM. Unfortunately, they were known to burst and burn at temperatures as low as 250 degrees centigrade, and catastrophically at 400 degrees.[5]

Unlike conventional reactors, the Windscale piles were vertical and cartridges were loaded into channels in the graphite base from front to back. The piles usually operated at a temperature of about 120 degrees centigrade to maximise the production of plutonium in the irradiated uranium. The cartridges were then pushed out the back of the pile into waiting skips, stored under water and subsequently processed to extract the plutonium metal. The control rods, used to switch the plant on and off, generated nuclear heating in the pile to a temperature of about 250 degrees during the procedure designed to rid the graphite of pockets of accumulated Wigner energy.

By October 1957, fourteen such procedures, with varying degrees of success, had been carried out on the two piles; eight on Pile No. 1 and six on the second pile. There was a risk of fire if the overall reactor temperature rose too quickly or became too high or if there was a cartridge burst, especially of the AM type, during the process. If the annealing temperature was too low, pockets of Wigner energy would remain trapped in the pile. If all the trapped Wigner energy was not released, as foreshadowed in the spontaneous heating incidents in both piles in 1952, the reactor could potentially overheat and catch fire without warning. The operators at Windscale may not have been aware of it, but the piles were an accident waiting to happen. Edward Teller's prediction about graphite reactors was set to come true.

* * *

On Sunday, 6 October 1957, Pile No. 1 was shut down for routine maintenance. The plan was to carry out a Wigner release and then discharge and refuel one zone of the pile after it had cooled down. Britain, Ireland and much of northern Europe were caught in the grip of an Asian 'flu epidemic and the maintenance schedule for Pile No. 1 was left in the hands of an unusually small number of staff. Apart from one channel where the cartridges had been removed, the pile remained fully loaded. The following morning, Monday, 7 October, the cooling fans were shut off and nuclear heating induced in the pile by manipulating the control rods. By midnight, the desired temperature of 250 degrees centigrade was reached and at 4 a.m. on Tuesday morning the nuclear heating was suspended and the pile shut down.

By now, in theory, the Wigner energy release should provide sufficient heat in its own right to generate a uniform Wigner discharge across the pile, but at 9 a.m. the operators noticed that temperatures in the core were either static or falling.

All the senior staff had gone home. An incomplete or failed discharge would increase the fire risk once the pile returned to normal operation, so a decision was taken to commence a second nuclear heating at about 11 a.m., applied until about 7.30 p.m. that evening.

At mid-afternoon on Wednesday sections of the pile were showing temperatures up to 415 degrees centigrade. Although temperatures returned to acceptable levels when cooling fans were turned on at 10.15 p.m., at midnight they began to rise again. A spike in atmospheric radioactivity levels recorded at the site meteorological station was attributed to Pile No. 2, where a faulty gauge had been detected the previous day.

Blowers were used to cool the pile down on Thursday morning. In the afternoon, temperatures again began rising inexorably. The engineer in charge suspected a burst cartridge. The Works manager, Hugh Gethin Davey, was called in and plans were laid to push the fuel cartridges from the channel where it was believed the burst cartridge was located.

Thirty years later, Arthur Wilson, the instrument technician who claimed to have first discovered the fire, told the *Guardian* his early warnings to management that the pile was on fire were ignored. His instrumentation – the uranium thermocouples that measured the temperature in the pile – had melted, he said.

'It just kept getting hotter. I telephoned someone who said not to worry, it will soon cool down,' he was reported as saying. 'I was not happy so I went up to an observation window that cut through the twelve foot of concrete [biological shielding] to have a look at the discharge face.' 'There were little jets of flame coming out of it. It was like the discharge out of the back of a jet engine.'

Wilson claimed that when he reported what he had seen, he was told not to be 'so bloody silly' and the fire was allowed to burn for a further six hours with nothing much being done about it.[6]

At 2 p.m. on Thursday, air samples from the on-site met station and from a building half a mile away showed high levels of atmospheric contamination. The only Health Physics van at Windscale was dispatched south along the coast to collect air samples and measure atmospheric gamma radiation. A second Health Physics van arriving later that afternoon was sent to take readings north of the site. The mobile results convinced health and safety staff that environmental radiation levels did not justify local evacuation.

Throughout Thursday evening and the early hours of Friday morning, production workers attempted to create a fire break around the blazing core by clearing adjoining channels of their fuel elements and, later, from the burning channels. The workers used whatever steel rods were available, including scaffolding poles brought across from Calder Hall. The pole tips melted in the core. Cooling fans were kept on to maintain some level of tolerable working conditions for the operators.

Davey's deputy, the Irishman Tom Tuohy, had arrived on the site at 7 p.m. on Thursday evening, called in from home where he was nursing his wife and children, who were all ill with 'flu. Tuohy also went to the observation holes and by 11.30 p.m. could see blue flames at the back of the pile. By now, the temperature in parts of the core had reached 1,200 degrees centigrade. The fire was spreading and growing in intensity.

Windscale was then the only nuclear site in Britain with a designated local emergency plan. At 1 a.m. on Friday morning, the Chief Constable of Cumberland was advised of a possible district emergency. A fleet of buses assembled to evacuate local residents. At the site, workers were warned to stay indoors and wear face masks. By 4 a.m. Tuohy feared the biological shield on which he was standing was about to collapse. A tank of carbon dioxide brought from the Calder Hall site was injected into the core, to no effect. The Windscale Works fire brigade was placed on stand-by.

There were no fire hydrants adjacent to the Windscale piles. Water had always been kept at a distance because of the presumed danger of a criticality incident if it came into contact with plutonium. Pouring water on burning graphite risked a hydrogen explosion. But by early Friday morning, there was no remaining alternative. Touhy turned on the water. At first, it was dribbled onto the burning core in a trickle.

The water pressure was gradually increased, but failed to have much discernible effect until the cooling fans were shut down at 11 a.m. The critical moment when an explosion might have scattered the entire pile and its contents over the whole of West Cumberland came and went. By midday, the fire was on its way to being finally extinguished.

The UKAEA Director of Operations, K.B. Ross, who had been on a visit to Windscale and was present throughout the crisis, belatedly sent a message to the UKAEA Chairman, Sir Edwin Plowden:

> Windscale pile No. 1 found to be on fire in middle of lattice at 4.30pm yesterday during Wigner release. Position been held all night but fire still fierce. Emission has not been very serious and hope continue to hold this. Are now injecting water above fire and are watching results. Do not require help at present.[7]

Plowden immediately wrote to the Prime Minister: 'I have to report to you a serious incident at Windscale ...' He attached a brief memorandum setting out the facts as he knew them.[8] What had begun as a local emergency in Windscale was now a national political issue.

* * *

At the site, health and safety personnel struggled to interpret the risk to the local population from the radioactivity that had continuously streamed out of the pile stack throughout the fire. A biological sampling programme was initiated on Friday morning. Local milk samples from Friday afternoon showed increasing levels of iodine-131, a short-lived isotope that lodges in the thyroid and presents a particular risk to children and pregnant women.

In 1957, the International Commission on Radiological Protection (ICRP) radiation guidelines applied only to workplace exposure. From Thursday evening the health and safety team at the Industrial Group headquarters at Risley, including the Chief Medical Officer, Andrew McLean, and health physicist John Dunster, were anxiously enquiring about possible public health hazards. Help was not needed, they were told.

By Saturday evening, the scientists recommended an eighty square mile milk ban around Windscale, later extended to 200 square miles, in a bid to minimise future cases of thyroid cancer arising from exposure to iodine 131.[9] Milk restrictions on Lancashire farmers were lifted on 1 November, but the ban continued in Cumberland for a further three weeks. Almost half a million gallons of milk were disposed of throughout the remaining days of October and a further 235,000 gallons in November. Farmers were compensated at a rate of one shilling and seven pence to the gallon.

Atmospheric contamination peaked in two major releases of radioactivity; the first late on Thursday night as the fire raged and spread throughout the reactor core; the second on Friday when the release of water onto the pile created a pall of black smoke and ash that could be seen from Millom, a town twelve miles away to the south of Windscale.

As with all major industrial accidents, there was no immediate certainty as to the extent of the environmental impact or where the radioactive plume would go. The only weather vane on the site was positioned at too low a level to accurately determine prevailing wind conditions. But the response of the Windscale management, initially to localise the crisis and then stolidly insist its impact was negligible, was symptomatic of a cover-up culture and the type of willful understatement that was becoming endemic in the British nuclear industry, of which Ross's 'nothing much to worry about' note to the Authority chairman was indicative.

The Authority's first press release on the accident issued on Friday afternoon – the same memorandum that had been furnished to the Prime Minister and the Ministry of Agriculture – suggested the Windscale filters had retained any radioactive release and 'there is no evidence of there being any hazard to the public'.

A UKAEA spokesman, based in London, provided his own interpretation of the events that had taken place at Windscale during the previous forty-eight hours in a quote splashed across the front pages of British and international media the following morning.

'There was no explosion,' the spokesman said. 'There was not a fire in the normal sense of the word. There was not a large amount of radioactivity released. The amount was not hazardous and in fact it was carried out to sea by the wind.'[10]

A plume carrying over 20,000 curies of iodine 131 was already drifting across south Lancashire and Yorkshire. A cold front moved in from the west over England early on Friday morning and the radioactive cloud was pushed south-west, reaching London at about 6 p.m. on Friday, 11 October, where its presence was detected on a monitoring filter at the Kodak factory in Harrow. Later it was detected in Belgium and within twenty-four hours at Frankfurt in Germany. Filter records showed it had extended over Holland by Saturday 13

October and within a few days traces were picked up in the south of Sweden.[11]

No warnings were given to other countries of the potential radiation hazard being carried on the wind. At the time no international protocol or practice existed for alerting other states to radiological incidents, but then it was hardly common practice either to set reactors ablaze and disgorge massive quantities of radiation to the environment. As well as 27,000 curies of airborne contamination, a further estimated 17,000 curies were discharged to sea from the highly irradiated water used to quench the fire. Later studies showed that only 40 per cent of the radioactive release, or about 4,000 curies, was trapped by 'Cockcroft's follies', the Windscale filters.[12]

Italy, about to sign off on a purchase agreement for a magnox plant for its Latina power station, was the first European country, on Saturday, 12 October, to seek reassurances from the British government.[13] The Italians were closely followed by the Belgians, whose own research reactor, developed with British assistance, was a smaller-scale model of the Windscale piles. When news of the fire was reported in the US, the acting chairman of the United States Atomic Energy Commission wrote to Plowden immediately, offering every assistance. The French made no direct enquiries, possibly because they feared they might be refused information,[14] but months later in February 1958 UKAEA officials and scientists offered them a declassified briefing on the incident.

There is no official record of requests for information made to the British embassy in Dublin or directly to London by the Irish government. The Windscale accident was not mentioned at any of the cabinet meetings throughout the month of October, although the *Irish Times* had reported growing fears of contamination in Ireland after the milk ban was introduced in the Windscale area on Monday, 14 October.

'There has been no change in the amount of radioactive material in the air over Ireland,' the paper reported in its city edition the next day. 'The amount of radioactive substance in the air over Dublin is measured carefully daily at the Dublin Institute of Advanced Studies in Merrion Square,[15] and an official of the school reported yesterday that there was no change in the rate.'

* * *

Given the magnitude of the incident, the British authorities were getting off relatively lightly despite their historic and deliberate policy, which had always made Hinton and Cockcroft uneasy, of keeping the public in ignorance about the less prestigious and hazardous aspects of atomic energy. The only real threat to the UK government's policy objectives was in the United States, where reports of adverse comments by one US senator, that the immediate area around Windscale would remain uninhabitable for two hundred years, generated a diplomatic frisson.

Predictably, the British tabloid press focused on the human interest angle – such as the *Daily Herald*'s photographs and reports on the case of Windscale operator Stan Ritson, 'The man with the radioactive hair', who had suffered contamination in the course of the incident. As the days passed, such stories were harder to come by. Within forty-eight hours, the *Liverpool Daily Post* would report that 'the shroud of secrecy which blanketed the whole emergency was well-maintained by the staff at the factory. Workers leaving the plant last evening refused to talk to reporters.'

Such inhibition was not confined to the ordinary operators. Hugh Gethin Davey was expressly prohibited by the Authority from making any public statement or engaging with an increasingly agitated and concerned local community. Radioactivity detected in local mineshafts and the fears expressed by farmers over the market value of west Cumbrian livestock were widely reported in British and some Irish newspapers.[16] Eventually, as the clamour of local concern grew louder, the UKAEA gave permission to Davey to speak at local public meetings.

'Had Mr. H.G. Davey been allowed to speak as freely last Friday as he did at a conference at Windscale yesterday [Wednesday, 23 October] a lot of misunderstanding and apprehension would have been avoided,' the *Whitehaven News* noted angrily.

Dr Frank Leslie, a Windscale scientist, in a letter to the *Manchester Guardian* published on 15 October, deplored the lack of public warning to local people as the reactor fire had developed. Leslie had taken background radiation readings in the garden of his Seascale home on Thursday 10 October, recording levels he considered sufficiently high to warrant advising the local population to take protective action and remain indoors. The Prime Minister was infuriated by Leslie's intervention, but stopped short of preventing him from giving an interview on television.[17] In Parliament, Macmillan also responded testily to questions from the local MP for Copeland, Frank Anderson, and other MPs who raised concerns about the public health impact of the accident.

The UK broadsheets rapidly turned their attention from the facts of the fire to the enquiry established by the government on 15 October to investigate it, with *The Economist* urging any such enquiry should be independent of the Authority. Most of the broadsheets were severely critical of the enquiry's composition – a supposedly reluctant Sir William Penney in the Chair; Dr BJF Schonland, Deputy Director of the UKAEA, and two academic engineers,[18] both of whom regularly worked as consultants to the Authority – and that its proceedings were to be conducted entirely in secret.

The Penney Report, compiled in ten days and presented to the UKAEA and the Prime Minister on 26 October, would, when eventually published in the truncated form of a White Paper, lay the blame for the fire squarely on operator error and the decision to undertake the second nuclear heating on Tuesday, 8 October.

The Authority board and the Ministry of Defence favoured publication of the complete report, but the Prime Minister demurred. Writing in his diary on 30 October, Macmillan noted: 'The publication of this report, as it stands, might put in jeopardy our chance of getting Congress to agree to the President's proposal.'[19]

* * *

Oblivious to such niceties of international diplomacy, management and workers at Windscale were battling to make the devastated reactor safe. The clean-up, over a period of several weeks, involved removing as many cartridges as possible from the damaged core and reprocessing them; sealing the concrete biological shield and boarding up the top of the chimney to prevent any further radiation releases as well as dealing with large volumes of contaminated water from the fire. Of 180 tonnes of fuel in the reactor, 158 were recovered. Of the remaining twenty-two, five were estimated to have been incinerated in the fire, leaving fifteen to seventeen tonnes of uranium twisted and melting in the still-smouldering damaged channels of the core.

Many of the general operators at Windscale were grounded due to exposure to high doses of radiation in the first phase of the clean-up, and about fifty volunteers were recruited from the UKAEA Capenhurst plant[20] to provide additional manpower. Many years later, one of the Capenhurst crew, Les Jenkins, alleged in the media that health and safety standards were fatally compromised in the rush to render the reactor safe.

'They just didn't give a toss for health and safety,' he said. 'Their attitude was: Let's get this bloody pile safe and then we'll worry about the consequences.'[21]

Jenkins claimed film badges and dosimeters used to measure personal levels of contamination were thrown around carelessly on floors and windowsills and the results not properly recorded at the height of the panic-stricken clean-up. Embittered by a £30,000 settlement paid out by BNFL under a 1983 corporate compensation scheme the company had agreed with its trade unions, several years after he had been diagnosed with cancer, Jenkins was hardly an impartial witness.

But his personal story remains part of the most enduring legacy of the Windscale fire: the controversy generated over the impact of large-scale releases of radioactivity on human health and the number of deaths and injury and damage to the human gene pool caused in the long term. It was a legacy that refused to go away and was revived at every point of controversy surrounding the nuclear industry for the next forty years.

* * *

In time, the consequences of Sellafield radiation, and the direct implications of the Windscale fire for the health and safety of Irish people living along the east coast, would become a central issue in Anglo-Irish relations. There was little indication of this imbroglio to come in 1957.

News coverage in all the main dailies was impartial and focused mainly on the implications of the disaster for West Cumbrian farmers and miners and the loss of plutonium production for the British military programme. Only the *Irish Press* delved in any detail into the possible mutagenic effects of radiation poisoning in a series of articles, 'Threats of Atomic Radiation', by the Very Rev. Dr P.J. McLaughlin, published over three editions of the paper up to the end of the month.[22] Under a banner headline 'What is radiation doing to your Children?', Dr McLaughlin claimed that 100,000 'genetic deaths' could already be attributed to the US Bikini Bravo bomb test of 1954, many more than directly caused by the bomb blasts of Hiroshima and Nagasaki, and that as such it was 'one of the most remarkable ironies of history'.

'Damage is cumulative from one generation to another ...' he wrote. 'The long term result is thus bound to be degeneracy and final extinction of the irradiated race.'[23]

With somewhat less hyperbole, exaggeration and inaccuracy, the Rev. McLaughlin noted that build-up of radioactive wastes and possibility of nuclear accidents 'are of international concern' and safeguards against their effects 'cannot be confined to national borders'.

In a foretaste of the moral supremacy which would distinguish so much of future Irish debate on nuclear issues, he accused the 'so-called progressive or advanced and wealthy countries' of taking a 'too-lighthearted attitude' towards the 'hazards of radiation'. 'Smaller and poorer countries,' he suggested, 'have been more prudent in their radiological practices.'[24]

To what extent the Rev. McLaughlin's views, or the *Irish Press* editorial decision to publish his series of articles, were representative of the Irish establishment view at the time remains unclear. The *Irish Press* was widely regarded as De Valera and Fianna Fáil's propaganda sheet and regularly lambasted as such by the main opposition parties, Fine Gael and Labour. At a national political level, no parliamentary questions were tabled about the Windscale fire or any fallout dangers to Ireland when the Dáil resumed after its summer recess in October 1957, nor did the Windscale fire even rate a parliamentary mention for many years thereafter.[25]

In Britain, the publication of the White Paper on 8 November 1957, together with a parliamentary statement by the Prime Minister, temporarily mollified the liberal media. In its next edition, the *New Scientist*, for example, abandoned a previously critical stance to note that 'the official Windscale Report offers so frank and satisfying an explanation that the public is enabled to view the happenings of 10 October in proper perspective'.[26]

A month after the fire, management of the Rowntree factory at Egremont, a few miles from Sellafield, sought compensation from the UKAEA for ninety tonnes of chocolate crumb, manufactured between 11 and 15 October 1957. Rowntrees accepted the chocolate 'did not represent a radioactive hazard', but for public relations reasons wanted to dispose of it. The UKAEA at first resisted the compensation claim, but finally agreed to pay the costs of destruction.[27] In Whitehaven itself lines were being drawn across the community. Hugh Gethin Davey later reflected that while local people and organisations demanded safety assurances, 'there was no deeplaid antagonism or resentment.'[28]

In an editorial on the fire, the local paper went a step further: 'There are some people especially in the Barrow area who want Windscale Works scrapped,' the *Whitehaven News* observed. 'Windscale has given work to thousands of West Cumbrians for the past ten years,' the paper remarked pointedly. 'Whitehaven College for Further Education or the new school at Seascale wouldn't exist without it.' It concluded defensively: 'We are proud of Windscale and Calder Hall and we in West Cumberland have little fear of the future.'

Only in July 1958, nine months after the fire, did the UKAEA formally announce the permanent closure of Windscale Pile No. 1;[29] the same month the UK and US finally signed off on a bilateral agreement on nuclear defence.

The Prime Minister's worst fears that the Windscale fire would jeopardise the successful conclusion of this long-sought Anglo-American defence agreement, which had motivated so much of the clampdown on information about the fire, had proved unfounded. Plowden effusively wrote his congratulations to the Prime Minister on this 'major achievement in the field of Anglo-American relations'. Macmillan responded with gratitude, 'to you and all those who helped you'.[30]

PART FOUR

THE RISE AND FALL OF CIVIL NUCLEAR POWER

1. Ernest and Freda Walton on their wedding day.

2. E.T.S. Walton as a small child with his sister Dorrie. Following his mother's death Walton spent much of his early childhood with relatives in Armagh.

3. Windscale site in the early 1950s, dominated by the Windscale piles and tall chimneys.

4. Aerial view of Sellafield in 2005.

 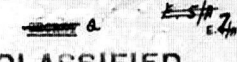

5. Lord Glentoran's enthusiasm for a nuclear plant in Northern Ireland was not universally shared in Whitehall.

6. Windscale under construction in 1950. Seven years later Pile No. 1 was destroyed by a fire and the two reactors were closed down.

7. Construction workers install steelwork for radiation filters on the Windscale chimney stacks in March 1950. The filters failed to protect the surrounding countryside from contamination.

8. The partly built Calder Hall reactor and cooling towers in the early 1950s. Calder Hall became the first commercial-scale nuclear power plant in the world. It was opened in 1956, and its operational life finally ended in March 2003.

9. The Windscale AGR, with its distinctive golf-ball shape, was the prototype for Britain's second-generation nuclear power plants and an iconic landmark of the Sellafield site. Decommissioning of WAGR has been continuing since March 1999.

10. The decommissioning of 5,702m of old pipelines used to discharge radioactive wastes and rainwater from the Sellafield site to the Irish Sea from the early 1950s was finally completed in June 2006.

11. Des O'Malley, Minister for Energy, Industry and Commerce in the late 1970s, at the opening of a new factory in Ireland.

12. Ray Burke (left), long a central figure in Ireland's anti-Sellafield campaigns, pictured with Joe Jacob in 1997 (courtesy of the *Irish Times*).

13. NCNI shop stewards Howard Rooms, John Kane and Douggie McCartney with their local MP, Jack Cunningham (second left), on their way to the first of several meetings with Prime Minister Tony Blair in Downing Street to plead the Sellafield cause.

14. The Sellafield workers' 'Trust Us' campaign was underwhelmed by Bono's attempts to stop Thorp.

15. Inside the Sellafield MOX plant. The 2001 decision to grant permission to operate the plant sparked Ireland's international court action against the UK.

16. The Irish yacht *Spinner*, and Greenpeace inflatables, confront the nuclear cargo ship the *Pacific Pintail*, at the approach to the harbour in Barrow-in-Furness in September 2002.

17. Irish Green Party MEP Nuala Ahern (right foreground, holding banner) took a leading part in the Irish Sea flotilla protest.

18. Greenpeace ship *Rainbow Warrior* leads the protest of Irish yachts across the Irish Sea.

19. Irish yacht makes the message clear.

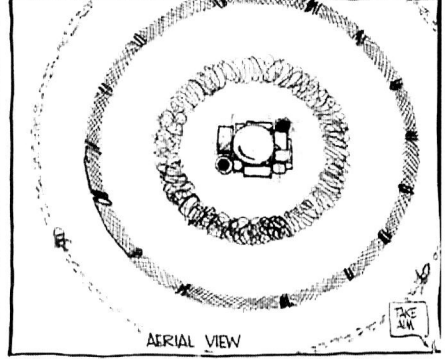

20. Cartoonist Martyn Turner's take on Sellafield's defence against terrorist attacks.

21. Labour Party Leader Ruairi Quinn in 2001 and party colleagues Emmet Stagg and Mary Upton take their skull and crossbones protest to the British Embassy in Merrion Road (courtesy of the *Irish Times*).

22. The Green Party's Trevor Sargent, with colleagues Ciaran Cuffe (left) and Eamon Ryan, planted paper windmills on the beach at the Poolbeg power station in Dublin to commemorate the twentieth anniversary of Chernobyl (courtesy of the *Irish Times*).

10

Ireland's Opportunity

'The wiser course is to look realistically at the world in which this nation has now to make its way.'

Seán Lemass, 1965

The image of the Sacred Heart, illuminated by a tiny, flickering red electric light, courtesy of the rural electrification scheme, adorned the kitchen walls of many Catholic homes in mid-twentieth-century Ireland, complemented by a statue of the Child of Prague, or the Virgin Mary, on a shelf or the mantelpiece over the fire.[1] Within the hollow centre of these gaudily, if often inexpertly, decorated plaster icons, householders stuffed their spare cash or concealed other official detritus of their lives.

These household shrines unwittingly provided a fitting symbol for a state whose outward displays of ostentatious piety and righteousness served only to mask the social and economic problems massing at its centre. As yet unacknowledged, the Sinn Féin nationalist revolution was failing and, unreformed, would continue to fail future generations of Irish people. Ordinary people, as Sean Lemass presciently observed in 1954, were losing confidence in the future of the Irish state itself.[2]

When the crunch came in 1956, some politicians blamed the availability of cheap American loans under the Marshall Plan for having encouraged profligate spending by government departments. Others pointed to the international environment, especially in the aftermath of the Suez crisis, as the root cause of Ireland's economic woes. Everything and everyone else was wrong, except the nationalist model for independence of which they themselves had been the embodiment since the foundation of the state.

In the seemingly never-ending parliamentary debates on the crisis, one Dáil deputy came uncomfortably close to a mark that the leadership of the main political parties would perforce deny:

The facts are that we are ineradicably linked up with Great Britain,' the deputy said. 'Our parliamentary system is based largely on the British Parliament; our way of life is much the same; our legislation follows the lead given by Great Britain and all our social legislation follows the British plan. That is because of our proximity to Great Britain and because we have to depend on Britain as our major market for our exportable surplus and for the supply of almost all our raw materials to keep industry going in this country.[3]

Economically and socially, no escape route from dependence on Britain had ever been successfully charted. It might all have been different, as the deputy said, if Ireland was an island 500 miles further west in the Atlantic.

The idyll of a self-sufficient, agriculturally led, Catholic enclave in the modern world, so beloved of De Valera and several of his contemporaries from the civil war political generation,[4] clashed with majority experience of life's ignominious daily grind. Irish independence was a political fact. But in many other respects, the country was wearing a coat that simply didn't fit.

The search for a distinctive post-colonial identity through revival of the Irish language was turning to cultural embarrassment as the number of Irish speakers diminished rather than increased. Successive Fianna Fáil and Fine Gael-led inter-party governments claimed it was impossible to provide an accurate tally of the real level of permanent emigration. The traditional safety valve of mass emigration, estimated by some to have reached 80,000 per annum during the bleakest years of the 1950s, was belatedly recognised as fatal to the country's future potential. In education, less than a quarter of second-level students progressed to university. The numbers who left formal education behind them at primary level remained much the same as at the birth of the state.

Any and all proposals for even the most modest educational reforms were routinely resisted by vested interests of church and state.[5] The same elites coalesced in strident opposition to progressive health care proposals, the most notorious example of which was the Mother and Child scheme sponsored by Minister for Health, Dr Noel Browne, whose forced resignation in April 1951 precipitated the fall of Costello's first inter-party government. As for the care and protection of children, and the treatment meted out to the less fortunate in Irish society in industrial schools, orphanages and other centres of care, only several decades later would scandals of emotional, physical and sexual abuse, and the corruption and abuse of power and privilege which sustained them, emerge to the cold light of public scrutiny.

Nationalist self-delusion had brought the Irish state close to self-destruction by the time De Valera shunted off the centre of the political stage in mid-1959 to spend the remainder of his political career, and most of the rest of his life, as Ireland's President.[6] As Dr Noel Browne declaimed after the 1957 election: 'We all know that Ireland cannot afford even one more mistake ... it is not just the Fianna Fáil Party alone which is on trial for the next five years; it is democracy itself.'[7] With De Valera's departure, Fianna Fáil, which had taken over from Costello's second inter-party government in 1957, came under the control of the ultimate political pragmatist and long-time leader in waiting, Sean Lemass.[8]

Supported by far-sighted civil servants, particularly T.K. Whitaker, architect of Ireland's first five-year economic plan in 1958, and surrounded by ministers predominantly drawn from the post-civil war generation, Lemass progressively deconstructed the old model of the nationalist state.

Agriculture was replaced as the driving force of Irish economic policy by the creation of sustainable, productive, industrial employment, with foreign direct investment selected as the preferred engine of growth. Looking to a future in Europe, Lemass set about dismantling the protectionist edifice he himself had been mainly responsible for erecting in the 1930s. In the seven years before his retirement from public life in 1966, Lemass wrought a transformation in economic outlook, educational reform and in Irish external relations, including with Britain, Northern Ireland, and Europe, from which there could be no going back.

A month after his appointment as Taoiseach, Lemass began talks with the British government on a free trade agreement. This first attempt foundered but the initiative was revived in 1963, culminating in a permanent Free Trade Treaty with the UK, signed in London on 14 December 1965. This permanent treaty was envisaged by Lemass as a temporary arrangement, another plank in his strategy to ease Ireland's eventual accession to the European Economic Community, which he hoped might be realised by 1970. The European project had been temporarily derailed in January 1963 by French opposition to British entry. As Irish officials and politicians would later freely acknowledge, Brussels was not prepared to entertain an Irish application, separate to Britain, for entry to the community.

An editorial comment in *The Times* of London suggesting the Free Trade Agreement signalled 'the end of the philosophy of Sinn Féin' immediately drew fire from T.F. O'Higgins, one of the grand old men of Fine Gael.

No revisionist questioning of Ireland's Sinn Féin route to independence would be tolerated. 'For us Sinn Féin has meant the coming of a change, which made it possible for an Irish Government to go to London and there freely to treat with the Government of Britain,' O'Higgins retorted.

'It would be unthinkable in normal circumstances for a country with [our] background and position, with the fears and susceptibilities we have had for very good reason, to turn back the pages of history and to make ourselves, as we had been, the larder of England,' he stated. But, as he acknowledged, 'these are not ordinary times.'[9]

Breaking new ground in the Republic's policy towards Northern Ireland, Lemass accepted an invitation from Captain Terence O'Neill early in 1965 to a meeting in Stormont followed shortly by a visit to the southern capital by the Northern Premier. From this historic exercise of hearts and minds, any mention of political reunification was excluded.

'I have on many occasions urged the desirability, in the national interest, of achieving, between the two areas into which Ireland is now divided, the maximum possible measure of co-operation in practical matters of public concern and expressed the conviction that this was possible without sacrifice of principle,' Lemass reassured those among his parliamentary colleagues who raised concerns that political claims to the North might be diluted.[10]

Although ultimately confounded by events in Northern Irish politics, this breakthrough in North–South relations established new levels of political and economic co-operation, including in the area of energy policy.[11] Later, it emerged that the construction of a nuclear power station, possibly sited on Lough Neagh in Northern Ireland, to supply the future energy requirements for both parts of the island, was an item on the official agenda for talks between the Taoiseach and Captain O'Neill.

* * *

Daniel Dixon, the second Lord Glentoran of Antrim, whose family had been at the heart of commerce and unionist politics in the province for generations, was reckoned the nearest thing to royalty ever produced in Northern Ireland.[12] Glentoran, Northern Ireland Minister for Commerce from 1952 to 1964, spearheaded a campaign to bring nuclear power to Northern Ireland even before publication of the 1955 White Paper in the House of Commons set targets for the new civil nuclear programme.

Pressing for a place for Northern Ireland, Glentoran initially met the Lord President, Lord Salisbury, who gave him a sympathetic hearing, and the Chairman of the UKAEA, Sir Edwin Plowden, in mid-1954. Glentoran was encouraged by the 1955 government announcement that included Northern Ireland as a possible location for any future UKAEA projects.[13]

Northern Ireland's claim was boosted by the province's need for a new power station of at least 120 MW by 1963. Glentoran argued that coal was much more expensive in Northern Ireland than elsewhere in the UK. A nuclear power station would ease unemployment and boost the morale of the Northern Irish population. If a place for Northern Ireland could not be found in the first civil nuclear programme, then his government would settle in the interim for 'one or more plants, which may be expected to be required for the production or processing of fissile materials or other related purposes'.[14]

The UKAEA had just announced Dounreay, at the remote northern tip of Scotland, as the chosen site for a prototype fast-breeder reactor. Since Northern Ireland was closer to the north of England UKAEA establishments, and possibly even more easily accessible to them than Dounreay, Glentoran was playing a relatively strong hand.[15]

Throughout 1955, Glentoran and his department in Northern Ireland relentlessly pleaded their case in letters and at meetings with various ministries in London and the UKAEA. In a letter of 2 June 1955 to the Lord President,[16] Glentoran invited Sir Christopher Hinton to Northern Ireland to view suitable sites for a magnox power station or a UKAEA factory, an invitation received with little enthusiasm. 'Sir Christopher Hinton sees no purpose in Lord Glentoran's inviting him to Northern Ireland. He would only go if instructed to do so for reasons of policy,' the Authority's internal correspondence records.[17]

Government officials were equally reluctant to satisfy Glentoran's ambitions to site a nuclear establishment in Northern Ireland, and civil servants in Whitehall connived with their UKAEA colleagues to find ways 'to dispose of them without a meeting', before eventually bowing to the inevitable.

'I have an instinct against an idea of putting a second Springfields[18] in Northern Ireland, largely on security grounds. I feel that it would not be wise for us to have any large quantity of uranium out of the United Kingdom,' the Secretary of the UKAEA wrote to Plowden in July 1955. A report to the Personnel Security Committee of the UKAEA in 1955 was more direct, emphasising the 'threat of IRA raids across the Border' and 'the difficulty in ensuring that civilians employed in Northern Ireland Defence Establishments are politically reliable'.

Glentoran's advisors remained impervious to technical arguments disputing the suitability of Northern Ireland for nuclear power generation. In a letter to the Northern Ireland Department of Commerce in May 1955, the UKAEA advised 'that a nuclear power station, if it is to be economic must be run as a base load station generating not less than 100 MW for the whole twenty four hours a day. The existing night load in Northern Ireland is believed to amount to not more than 60 MW and even if it were to rise to as much as 100 MW, the economics of the nuclear power station would require the nightly closing down of all other stations in Northern Ireland.'[19]

J.A. Mc Keown, the lead official in Belfast, commissioned his own consultants, Kennedy & Donkin, to prepare figures. Writing to Sir John Maud in the Ministry of Fuel and Power in June 1955, the UKAEA's Sir Donald Perrott remarked that McKeown 'seems determined to ignore the advice given to him by the Lord President and by your Minister in July 1954 and March 1955 [and] no doubt feels it unnecessary to wait for his Consulting Engineers' report before making up his mind!'[20]

Kennedy & Donkin's report rejected the load factor argument, nor was Belfast easily persuaded that a nuclear power station would do little to ease Northern Ireland's unemployment. In advance of a meeting planned for mid November 1955, Plowden was advised to impress on Glentoran that the construction force for a nuclear power plant 'would probably have to be imported (possibly Irish Catholic labour) and if it remained in Northern Ireland the position there would be worsened'.

Towards the end of 1955, the UKAEA, searching for a site for a new uranium enrichment plant to manufacture reactor fuel, shortlisted a number of former Second World War airfields. The list included Maydown airfield, close to Londonderry in Northern Ireland.

Northern Ireland could not accommodate a uranium enrichment plant because it lacked sufficient electric power to supply it. Economically, it lacked the electricity demand to justify building a nuclear power plant. A reactor and a fuel plant would fit on the Maydown site. The fuel plant could draw on the reactor for its power supply with any surplus going to the Northern Irish grid for general distribution. The raw material for the plant, uranium hexafluoride, could easily be transported by sea from Springfields to Londonderry, with the end product shipped by return from Londonderry to Whitehaven. Potential employment amounted to as many as one thousand jobs.

Plowden instructed that the government of Northern Ireland was not to be informed that the UKAEA was even contemplating a site in Northern Ireland. Arranging a visit to reconnoitre the Maydown airfield while maintaining secrecy as to its purpose was complicated, but Plowden was adamant that the less divulged to the Northern Ireland Department of Commerce the better. As one Authority official noted: 'His fear, of course, is that once the Northern Ireland Government know that we are seeking a site, they will not rest until they have succeeded in getting us to build something there.'[21]

If the Republic was preoccupied with establishing a distinctive post-colonial identity, Northern Ireland's persistence in demanding a place in Britain's nuclear programme was symptomatic of the unionists' desire to demonstrate their political and economic integration with Britain. One-upmanship over the economically ailing Republic may have been another factor.

Glentoran's department published a White Paper on energy policy for Northern Ireland in 1956, centred round a proposal for a nuclear power station. The locations of the twelve proposed nuclear power plants in Britain were not yet finally settled. The Northern Irish authorities did not expect to start work on the project until at least 1960.

Glentoran's public statements may have caused alarm within the UKAEA and at Whitehall, but they attracted more favourable interest in Dublin. Liam Cosgrave, the Republic's Minister for External Affairs, publicly welcomed the northern announcement and hoped any nuclear project in the North would have an all-Ireland dimension.

Glentoran and his fellow unionists in Belfast were hardly impressed that their southern counterparts saw an opportunity in Northern Ireland's nuclear prospects to promote Irish unity, a theme that would thread its way consistently through the fabric of all Irish political debates on nuclear energy up to the end of the 1970s. When asked in the Dáil in July 1956 if he had 'received any acknowledgments from the Six-County Government in reply to his expressed desire for co-operation in regard to the peaceful use in Ireland of atomic energy', Cosgrave could only reply, in a word: 'No.'[22] Two years later, responding to a question from a Nationalist Party MP at Stormont, on whether consultations had taken place with the Dublin government on public safety issues arising from the disposal of atomic waste from a Northern-based reactor, Lord Glentoran stated that the need for special consultation with other governments was not apparent. It would, in any case, be a question for the government of the UK, he said.[23]

* * *

Liam Cosgrave, son of the state's founder, William T., became leader of Fine Gael in 1965 and Taoiseach eight years later. Cosgrave was among the strongest political advocates of nuclear technology in Ireland throughout the 1950s. Cosgrave's earlier plan for Ireland to gain a foothold in the atomic age through an 'Atoms for Peace' project was dashed when De Valera's government rejected the Irish Atomic Energy Commission's recommendation to buy a US reactor in 1958. But the government's decision had been influenced more by the economic crisis facing the country than any hostility to nuclear power *per se*.

Across the political spectrum, and among the confraternities of Irish engineers and scientists, the common view was that an Irish nuclear project was not just desirable, but inevitable. Thus, the American administration's offer in 1960 to gift training equipment to Irish universities was graciously welcomed on all sides. The inter-government agreement to supply the equipment, including a sub-critical training reactor ultimately destined for University College Cork, was signed in Dublin on 24 April. As described by Frank Aiken, Minister for External Affairs, this 'magnificent gift' was valued in total at £100,000. The only costs accruing to the Irish government were £5,000 for transport and insurance while in transit. In return, Irish universities would provide the USAEC with any research results obtained from its use. It was a very generous offer that 'enables us to take a big step now in preparing ourselves for the future,' Aiken told the Dáil.

'Atomic energy, while disastrous when used in connection with weapons of war, is one of the best tools man has ever possessed,' he said; sentiments that reverberated around the house. Nuclear energy was a 'great boon to mankind', the Labour Party observed, whilst Fine Gael's Declan Costello urged the government to publish their long-term proposals for the development of nuclear power in Ireland.[24]

Irish politicians, like their counterparts elsewhere at this time, preserved a sharp distinction between the civil and military uses of nuclear energy. Aiken saw no contradiction in putting forward proposals to control nuclear military proliferation, as he did at the UN in 1958, alongside enthusiastic endorsement of the use of nuclear energy for peaceful purposes at home. Aiken's UN campaign culminated in the Non-Proliferation Treaty of 1967,[25] to which Ireland was invited to become the first signatory in recognition of his efforts.

Public concern and opposition to hydrogen bombs and atmospheric weapons tests were as manifest in Ireland as anywhere else, as was the general feeling of helplessness about the threat of nuclear war between the superpowers, particularly at the height of the Cold War and the Cuban Missile Crisis in 1962. Ernest Walton had by then become lifetime President of the Irish branch of the Pugwash movement, the organisation devoted to worldwide nuclear disarmament founded in 1955 by Bertrand Russell, Einstein and former Manhattan scientists, including Joseph Rotblat. Shortly before his own death in 2005, Rotblat explained the impetus behind the Pugwash conferences, for which he and the Pugwash movement were jointly honoured with the Nobel Prize for Peace in 1995. 'We took action then because we felt that the world situation was entering a dangerous phase, in which extraordinary efforts were required to prevent a catastrophe,' Rotblat said. Asked by a reporter in 1945 for his reaction to the bombing of Hiroshima, Walton had replied similarly: 'It is to be hoped that if an enlightened human race does not prevent the use of these discoveries in war then a frightened one will.'[26]

By 1962, Ireland had long since joined the ranks of the frightened. Many in the post-war generation remembered as children, at home and in school, being led in earnest prayers for a peaceful resolution of the Cuban Missile Crisis to avoid what their elders envisaged as an imminent fiery end to the world. Following the Cuban crisis, the government issued a booklet, 'Bas/Beatha',[27] to every household advising people on what action to take in the event of a nuclear conflict. Each booklet came with a neatly punched hole in the top left-hand corner, so that householders might hang it in a place, conveniently accessible, if the alarm was raised.

From the late 1950s, the government faced an increasing barrage of questions from parliamentary opposition and the media on Ireland's preparedness for a nuclear emergency in the event of war and from Dr Noel Browne in particular on the health effects of fallout from nuclear testing and radiation protection for Irish workers in the health sector and in industry.[28] Windscale pollution had also begun to find its way onto the Irish political radar screen, to the extent that Aiken made enquiries of the British government in 1960 about the level of discharges, reported at 10,000 curies of radioactivity per week, into the Irish Sea. Aiken stated he was 'satisfied from the information I have obtained that the amount of radio-active waste material being now pumped into the Irish Sea does not at present constitute a danger to the health of the people'.[29]

As the Irish economy expanded and demand for electricity grew in the early 1960s, the Electricity Supply Board began to seriously evaluate nuclear power as a future energy source for Ireland. Under the second Programme for Economic Expansion, the government established a new Expert Committee, again including Professor Walton of Trinity, to consider the question. In 1966, the committee recommended the construction of a nuclear reactor in Ireland.

* * *

Radioactive discharges to the semi-enclosed Irish Sea had grown exponentially throughout the late 1950s and 1960s as the Windscale Works expanded; first to cope with the throughput of materials from the military reactors, Calder Hall and the four Chapelcross magnox reactors in Scotland, and later from the first civil nuclear power plants about to come on line at Berkeley and Bradwell in England, and Hunterston in Scotland.[30]

Windscale's first reprocessing plant, the primary chemical separation plant, known as B204, was designed and constructed without any prior pilot plant trials. The chemical separation plant operated reasonably efficiently except for highly radioactive sludges that accumulated at various points throughout the process machinery and had to be flushed out, and the contamination of walkways and platforms that occurred whenever samples were taken at different stages of the process. The risk of fire was ever present, as the chemicals, the butex and nitric acid used for reprocessing, reacted violently with one another. The final stage of the process, the plutonium purification plant, was more problematic, its operations characterised by mechanical faults, failures and leaks from pipes and valves. Plant operators, often working in a highly contaminated atmosphere for prolonged periods, wore full protective clothing and air masks.[31]

Spent fuel for military plutonium production became available from Calder Hall in 1956 and from the first of the Chapelcross reactors in 1958. At Calder Hall, the irradiated fuel was packed in heavily shielded flasks and moved across to the adjoining Windscale site. The fuel elements were unloaded into a concrete-lined pond and stored under water for about three months to allow unwanted fission products, such as iodine 131 with its half-life of eight days, to decay completely. The fuel was then stripped of its metal cladding. The process wound its way through a series of interconnected pipes and vessels in which the fuel was first dissolved in acid and the uranium, plutonium and waste products separated into different streams. At the final stage, solvents were used to extract uranium and plutonium. The plutonium was shipped to Aldermaston for use in the weapons programme and research projects. This first-generation reprocessing plant at Windscale provided both a laboratory and a learning curve for a technology that would come to dominate Sellafield's later commercial operations.

Reprocessing was the logical choice for the treatment of spent magnox fuel. It recovered valuable materials – weapons-grade plutonium, Pu 239, from the military reactors at Calder Hall and Chapelcross and a mix of military- and civil-grade plutonium from the civil magnox fleet. The wisdom of accumulating a stockpile of civil plutonium was never questioned. Most was earmarked for future use in fast-breeder reactors, of which a prototype was under construction at Dounreay in 1955. Reprocessing also created valuable commercial opportunities, not just in handling the spent fuel output of British power stations, but internationally. The technology was potentially attractive to natural resource fuel-starved countries such as Japan: the nuclear fuel cycle offered long-term security and stability of supply in an internationally volatile energy market.

The 1958 Mutual Defense Agreement (MDA) with the United States provided a further incentive to use reprocessing as a waste management technology. The MDA provided for barter arrangements with the US of UK plutonium, in exchange for enriched uranium to fuel Polaris submarine reactors and tritium for warheads. The advantage of Britain's civil reactors was that they were not part of either 'Atoms for Peace' or Euratom's non-military research programmes. British plutonium supplied under MDA barter arrangements could thus be used for either civil or military purposes without conflicting with other US international undertakings.[32]

In the early 1980s, a Central Electricity Generating Board (CEGB) scientist, Ross Hesketh, who had formerly worked at Dounreay on the fast breeder project, alleged that plutonium from the civil magnox reactors was diverted to the military plutonium stockpile in the UK and subsequently transferred to the US for use in weapons manufacture.[33] Figures released by the Clinton administration in the mid-

1990s, and later confirmed by the UK Ministry of Defence, showed that 5.4 tonnes of plutonium were transferred to the US from Aldermaston between 1960 and 1979 in barter transactions.

According to a US Department of Energy statement in 1997, some of the plutonium acquired under the barter arrangements up to 1964 was definitely used in weapons manufacture, a point the British authorities remained reluctant to concede.

'It is theoretically possible, but very unlikely, that some UK civil plutonium may have been transferred to the US and used in the US nuclear weapons programme before 1964 ... records do not exist to determine this with absolute certainty at this remove,' a Ministry of Defence statement said in 2000. Notwithstanding any deficiency in record-keeping in the early days of the programme, and the possibility that some records may have been destroyed, it seems more likely than not that a substantial portion of the plutonium used in barter exchanges up to 1971 came from the new civil reactors and was probably used in the manufacture of nuclear weapons in the US.[34]

A small plant at the north end of the Sellafield site was engaged in some unusual reprocessing: the separation of americium from plutonium warheads. Left to itself, Pu 239 decays into americium in about fifteen years, heating up the weapons' core and rendering them useless for military purposes. For this reason, warheads were stored separately from the rockets and specialist reprocessing was required periodically to remove the contaminating americium.[35]

Evidence of plutonium reprocessing is clear from the pattern of site discharges in the 1960s, when americium discharges peaked over plutonium for prolonged periods.[36] Thereafter, it is difficult to pinpoint when this activity was carried out, as new clean-up plant limited the discharge of both americium and plutonium.

* * *

In the late 1950s and 1960s, however, Windscale's main preoccupation was to resolve the technical challenges posed by expansion of civil nuclear power. A new reprocessing plant, B205, was built to cope with the existing throughput from the military reactors at Calder Hall and Chapelcross and the anticipated increase in volume from the new magnox civil stations. B205 was finally commissioned in mid-1964. Another new 'head end' plant, B30, equipped for the first stage of magnox reprocessing – chopping and separating the cladding from the fuel rods – and a large storage pond were also built. Looking forward to second-generation nuclear power plants, which would burn uranium oxide rather than magnox uranium metal fuels, the ten-storey Windscale reprocessing plant, B204, was decontaminated and converted into a head end plant for the new oxide fuels.

As well as contracts for the CEGB magnox stations, Windscale secured contracts to reprocess magnox fuel from Italy and Tokai-Mura in Japan, the only commercial reactors ever sold abroad by the UKAEA. Dounreay also set up a small reprocessing business, with contracts for spent fuel mainly from research reactors in Japan, Canada, Germany and Denmark. By 1965, the British government enjoyed sufficient confidence in the future of an internationally based reprocessing business in the UK to commercialise the fuel services division of the UKAEA. Five years later, in early 1971, Edward Heath's Conservative government dusted off a Bill left behind by Harold Wilson's Labour administration and broke up the UKAEA, establishing British Nuclear Fuels plc (BNFL) as an internationally trading, state-owned, commercial enterprise.

BNFL anticipated a bright future, including early privatisation.[37] But the nuclear industry was no longer immune from public scrutiny in a new age of politics dominated by a general public demand in the democracies of the West for greater accountability from their governments. This found expression across a wide range of issues, including the environment, in the rapid proliferation of non-governmental organisations and single-issue lobby groups. At Windscale, an overly rapid and ambitious expansion of operations was set to reap dividends, but not quite those anticipated by BNFL or its allies in Whitehall.

11

The Plutonium Economy

'Windscale – The world's nuclear dustbin.'
Daily Mirror, 1975

In an era when Britain's imperial status had all but vanished, its economy was struggling and neither hope nor glory were much in prospect, Calder Hall had evoked genuine national pride as the western world's first commercial-scale reactor, right across the political spectrum and amongst the general public. A uniquely British achievement of the atomic age, its primarily military purpose was conveniently ignored. Calder Hall was the standard bearer of Britain's civil nuclear programme.

A technology that Hinton and Cockcroft had believed should take decades to evolve and refine was scaled up for the delivery of ever larger reactor units as Britain's early civil nuclear programme rolled out.[1] The programme proved more expensive than the industry, policy makers or the generating companies and consumers, forced to pick up the tab, had expected.

Although they conformed to the same basic model, each of the magnox stations incorporated unique design features, arguably over-engineered, that pushed up capital costs. Costs were further exacerbated by frequent construction delays. In operation, unanticipated technical problems undermined generating efficiency and kept nuclear uncompetitive. Yet nuclear power was 'national policy, as laid down by the Government', the permanent secretary to the Department of Fuel told the Select Committee on Science and Technology in May 1962. The Central Electricity Generating Board, and its counterpart in Scotland, had to accept additional generating costs, estimated at £100m over conventional power stations in the programme's first five years, until such time as it became competitive, the permanent secretary said.

With traditional British stiff upper lip, the committee reported: 'The argument as to whether the taxpayer or the electricity consumer should bear the extra cost of the nuclear power programme should proceed from the basis that in ten or fifteen years' time nuclear power stations will be needed, and that just as present customers have benefited from technological advances in the past, so they should bear the cost of present advances.'[2]

Expansion of the first nuclear programme to 6,000 MW by 1965, announced by the government in 1956, was premised on a series of negatives, none of which transpired. Oil prices were expected to rise. There were fears that supplies from the Middle East, on which the Western world was becoming increasingly dependent, might be curtailed following the Suez debacle. The government was worried by the prospect of a shortfall in Britain's largest indigenous energy resource, coal. Electricity demand was expected to grow by at least 2 per cent per annum.

As it was, over the next two years electricity demand flattened to zero growth. Internationally, the price of oil fell as Middle Eastern and Libyan supplies to Europe and the US increased. In Britain, despite persistent rumblings of industrial unrest among the miners, coal production remained steady and technological innovations in coal-fired power plants ensured a competitive advantage over nuclear-generated electricity. In 1960 the ambitious programme for nuclear expansion was scaled back to 5,000 MW with a new target date of 1968 for completion of the first phase. Only nine of the projected twelve magnox stations were ever built.

Commercial exploitation of nuclear power was delayed in the United States until 1959, when the first commercial project underwritten by government guarantees, a 200 MW boiling water reactor (BWR), designed, equipped and built by General Electric, came on line in Chicago. Five years later, GE recorded a further triumph when it opened a 500 MW BWR at Oyster Creek in New Jersey, built entirely without government subsidy. The leading American development companies, Westinghouse and GE, had capitalised on their experience in building submarine reactors for the US navy in the design of water-cooled reactors for large-scale power generation. Light water reactors, BWR or pressurised water reactor (PWR) types, commonly regarded as safe, efficient and economical to run, and offered as 'turn key' projects to international customers, became the market leader in the western expansion of nuclear power in the late 1960s and 1970s. Logically enough, since it was an international commercial success, the

PWR became the prime target for anti-nuclear campaigners questioning the safety of nuclear power.

* * *

Unlike France, which made the American reactor design the basis for its expanded nuclear programme, Britain held out against the international trend in favour of the PWR. Dithering and vacillation characterised the debate about the scale of the programme or the selection of one reactor type over another in which the industry and successive British governments indulged in the 1960s and early 1970s.

A 33 MW output advanced gas-cooled reactor (AGR) prototype, a next-generation magnox, had been operating successfully at Windscale since December 1962. By 1964 the CEGB already had four AGRs on order. There was an implicit assumption that spent fuel from British second-generation reactors would, like magnox, undergo reprocessing at Windscale.

Expansion in reprocessing at home reinforced the potential of a new internationally trading industry in nuclear fuel services, justifying the separation of the UKAEA's fuel services divisions into the new BNFL. Such services would in time make BNFL one of Britain's largest industrial foreign currency earners. Indeed, by the late 1990s BNFL accounted for 17 per cent of UK service exports to Japan, the largest share of any individual UK company.[3] Apart from foreign currency, the attraction of an international fuel cycle business to successive governments lay in the chance to recoup decades of investment in Britain's nuclear project. Less critical analysis was applied either to the robustness of reprocessing technology or the competence of those operating it.

Reprocessing technology was designed to deal with military cycle uranium, in which the fuel cycle was short and the fuel was removed from the reactor after only a few months when its concentration of weapons-grade plutonium, Pu 239, was maximised. Reprocessing recovered between 0.1 per cent and 1 per cent plutonium, 96 per cent of the original uranium and 3 per cent unwanted highly radioactive fission products, designated as high-level waste. These 'high-level wastes', piped directly into storage tanks, were condensed by boiling under their own radioactive heat.

All of the magnox stations except for Oldbury and Wylfa, which were equipped with dry store facilities, were fitted with a standard pool to hold spent fuel. After resting in the pool for about 150 days, the fuel was transported to Windscale where it was again stored under water to await reprocessing.

Advances in fuel design, stimulated in part by the need to improve the efficiency of the new civil reactors, meant fuel could be kept for ever-lengthening periods in reactors. One unintended consequence was that extended neutron radiation of magnox fuel rods resulted in formation of new compounds, which were released into the high-level waste stream. If high-level waste containing these compounds was allowed to boil, the compounds created residues at the bottom of the waste storage tanks, and risked corroding them. With higher volumes of high-level waste putting pressure on tank space, BNFL urgently needed to construct new tanks, without which reprocessing could not continue. B205 was temporarily shut down in September 1972.

By the time B205 came back into operation the following summer, a backlog of magnox fuel had accumulated in the Windscale pond. The temporary storage ponds at power stations throughout England and Scotland were also filling up. In Berkeley, Britain's first civil magnox station, 9,000 fuel rods were stored in a temporary pool which had a nominal capacity for 2,000 rods.

Prolonged immersion of magnox fuel in water corroded its magnesium alloy cladding. The fuel deteriorated, making it more difficult to reprocess. As the backlog of magnox fuel at Windscale increased, and fuel sludge seeped through the disintegrating cladding, the Windscale pond became progressively more highly irradiated. Surface water discharges from the pond were heavily contaminated, especially with caesium 137 and plutonium. Radioactive discharges to the Irish Sea reached their highest point in the mid-1970s.

Reprocessing of oxide fuels from the second-generation nuclear plants also threw up some unexpected technical challenges. B204 had been converted to a head end plant for the treatment of oxide fuels, on which BNFL's international business ventures were based. Uranium oxide fuel was used in the US water-cooled reactors that had been sold throughout the world. It was also used in Britain's AGRs. This ceramic fuel, encased in stainless steel or a zirconium alloy, had a burn-up of several years before refuelling was required. When removed from the reactor, oxide fuel was ten times more radioactive than magnox fuels.

The higher reactor burn up increased the number and complexity of fissile products within the spent fuel, some of which failed to dissolve properly in nitric acid. Instead, they settled at the base of vessels at different stages of the process. The heat they produced evaporated all liquid surrounding them, leaving behind a crust of intensely hot granules. On 26 September 1973, operators in B204 prepared to process a new batch of oxide fuel. When new process liquids made contact with the heated granules in the pipework the result was a steam explosion which contaminated the entire building and the thirty-five workers within it. B204 was shut down.

* * *

The accident in B204 left BNFL with some 350 tonnes of oxide fuel from foreign customers and no plant in which to reprocess it. BNFL's annual report for 1974–5 downplayed the accident. The report simply noted that reprocessing oxide fuels was 'an exacting task'.

'Operations on the "head end" plant at Windscale for handling such fuels have had to be suspended until such time as plant modifications have been completed, but the experience gained is being applied to the new plants now being planned,' it said.

This obliquely referenced 'new plant' was a thermal oxide reprocessing plant, THORP, plans for which were first unveiled to BNFL's workforce in November 1974. At first the company considered building two Thorps, each capable of reprocessing 1000 tonnes of spent oxide fuel per annum. One Thorp would deal with British AGR spent fuel and was expected to be in operation by 1983. A second plant for international reprocessing contracts would come on line in 1986. By 1975, the company decided one Thorp would suffice.

BNFL was already engaged in preliminary discussions with the nine regional Japanese utilities. This initial sales drive netted potential contracts for 4,000 tonnes of Japanese spent fuel. The Japanese power industry was prepared to make a substantial pre-payment for fuel reprocessing services in the form of investment in the construction of Thorp. Further, they agreed to elastic contractual provisions that would permit BNFL to increase prices on existing contracts in a range of given circumstances. Between Japanese and European customers, BNFL anticipated securing international contracts to reprocess 6,000 tonnes of oxide fuel in Thorp's first ten years. With additional business from the British AGRs, this was more than enough to keep the new Thorp plant in business for a decade.

At the start of the 1970s, European fuel cycle companies feared American competition for international reprocessing contracts would dominate the market by 1980. In anticipation of this threat, Britain, France and Germany pooled resources in 1971 and formed a joint company, United Reprocessors, to exchange technical information on reprocessing. The participating companies agreed not to undercut one another in bids for international contracts. Like BNFL, the newly formed French fuel cycle company, Cogema, already had the benefit of military reprocessing experience at its plant at Cap de La Hague in Cherbourg. Germany was planning a reprocessing plant at Gorleben to service its own nuclear industry.

The American reprocessing competitor never materialised. A moratorium on the development of commercial reprocessing was declared, first by the outgoing President, Gerald Ford, and endorsed by his elected successor, President Jimmy Carter, in April 1976, as part of US policy on nuclear non-proliferation. The US administration would not countenance the development of a civil industry based on the extraction of plutonium, even if the ostensible purpose of creating a so-called plutonium economy was entirely peaceful and the plutonium was destined solely for use as fuel in fast-breeder reactors of the future. Although it was not acknowledged at the time, the dream of fast breeders as the ultimate fission reactor, the persistent promise of 'jam tomorrow', was beginning to wear thin and most countries, including Britain, would cease investment in their development by the late 1980s.

Since the enriched uranium used to fuel Japanese plants originated in the US, the movement of spent fuel and the ultimate repatriation of plutonium to Japan under the reprocessing contracts relied on US permission. The diplomatic wrangling that followed from the US moratorium on development of its own reprocessing industry was set to work out to the European companies' advantage. In security terms, the US seal of approval legitimated nuclear transports, making it more difficult for transit countries, especially those who hesitated to offend American opinion in any way, to object to them.

Only BNFL and Cogema were in a position to offer civil reprocessing services on the European and Asian markets by the start of the 1980s. If anything, a shortage of reprocessing capacity rather than a surplus looked imminent.[4] Reprocessing contracts were underwritten by international government accords. Domestic political commitment to the international reprocessing trade was essential to its success.

* * *

United under the banner of OPEC, the oil-producing nations of the Middle East had by 1970 belatedly come to recognise the political power of the cheap black gold on which the Western world relied for its prosperity. Inspired by Libya's example under Colonel Ghaddafi[5] of forcing price increases from the international oil companies exploiting his country's natural resources, and partly in revenge for American and European governments' support of Israel in the Middle Eastern wars of 1967 and 1973, OPEC turned oil into a political weapon. Overnight, on October 16 1973, the cartel trebled the price of crude oil on the world market to $11 a barrel.

Less cushioned against a massive increase in energy prices than they would learn to be by the end of the century, the economies of Western Europe and the US were driven into recession, fuelled by rampant inflation as oil prices continued to rise in the environment of general uncertainty. In the minds of energy policy-makers, security of supply was suddenly elevated from what had previously been thought of as theoretically desirable to a national priority.

Unlike France and Japan, the more chequered political culture of Britain coupled with intense rivalry between entrenched vested interests in the energy sector precluded any national consensus on the expansion of nuclear power as a long-term solution to security of supply concerns. For a start, the UK government and its policy advisors were still, as they had been throughout the 1960s, struggling with a final decision on a second-generation reactor.

It was obvious that the four AGRs already built or under construction were way over budget and had extensive engineering problems but politically, a marked reluctance to switch to the American water-cooled design persisted. Britain was preparing to become an oil-producing nation in its own right. The coal unions' simmering unrest had boiled over in a major strike by late 1973. Edward Heath's government, forced into an election to confront the miners, called for 28 February 1974, instead became its political victim and Harold Wilson's Labour Party was returned to power.

Wilson appointed Tony Benn as Secretary of State for Energy in April 1975. Enthusiasm for nuclear power, which Benn had exuded in the 1960s, was by now replaced with scepticism and paranoia about the nuclear industry that, in his terms, seemed to want a nuclear rod in every pot and a power station in every driveway.

For ideological reasons, Benn was also adamantly opposed to the PWR design, recording in his diary in 1976: 'I shall fight like a tiger against the American light water reactor.'[6]

In January 1978 Benn announced that two further AGRs would be constructed. Any decision on PWRs could be postponed to the 1980s, he said.

In the whirlwind of political argument about reactor choice for Britain, BNFL's audacious plan for an international fuel services industry involving the importation of 4,000 tonnes of spent fuel from Japan at first passed largely unnoticed by the public and the press. The financial media and trade journals wrote up the Thorp project in tones of general approval, but elsewhere it was ignored.

* * *

Environmental organisations such as Greenpeace, founded in 1971, and its breakaway organisation, Sea Shepherd, in 1977, were more concerned with species protection and ecological conservation than with energy issues in their early days. The environmental movement was at first more inclined to train its energy guns, so to speak, on ecologically suspect hydro-electric schemes or dirty coal plants rather than civil nuclear power, which they accepted as 'clean'.[7] But as locally based protest groups, especially in the US in the late 1960s, began, on grounds of safety, to challenge the location of nuclear power stations in their communities, the broader movement started to take note.

Friends of the Earth was founded in 1969 in the United States. A breakaway group from a much older environmental organisation, the Sierra Club, it had established its credentials as a worldwide federation of anti-nuclear lobby groups by the mid-1970s. Friends of the Earth UK was busily tracking developments on the Thorp project. The organisation published its own four-page leaflet, *Nuclear Times*, in May 1975. BNFL's Thorp plan, it claimed, would turn Windscale 'into one of the world's main radioactive dustbins'.

A few months later, Friends of the Earth spokesman Walt Patterson briefed a *Daily Mirror* reporter on BNFL's reprocessing plans.[8] He was duly obliged with a banner headline in the newspaper on 21 October. Its front page story 'Plan to make Britain the world's nuclear dustbin' and its call on the British public to 'Sign here for Japan's Atom Junk' generated a national furore, prompting the Minister for Energy, Tony Benn, to immediately call for a public debate on overseas reprocessing and BNFL to wake up to the reality that its plans for the future were far from secure.

* * *

The nearest Northern Ireland possibly ever came to realising Glentoran's dream of a nuclear-powered future was in the 1956 proposed expansion of the UK programme. By the time the last of the magnox fleet, the Wylfa power station at Anglesea in Wales, several years behind schedule, began generating electricity in 1971, Northern Ireland had politically imploded under the weight of its sectarian contradictions and was already in the grip of the beginnings of a long and bitter sectarian war.

Northern Ireland threatened to destabilise politics on the whole island and came close to doing so at the end of the 1960s and later as the Provisional IRA campaign of violence to expel the British from the province intensified.[9] For their part, bedrock unionists responded to any initiative for reform with intransigence; reaching its apogee in May 1974 with the loyalist workers' strike that brought down the first power-sharing experiment of moderate unionists and nationalists, the Sunningdale Executive.

Northern Ireland forced Anglo-Irish relations out of the bunker of traditional post-colonial resentment onto a plane of recognition of a mutual interest in not having Northern Ireland degenerate into anarchy, or the whole island into civil war. The British embassy in Dublin had been burned to the ground on 3 February 1972 following the 'Bloody Sunday' atrocity at a civil rights march in Derry the previous Sunday, when thirteen marchers were shot dead by the British Parachute regiment. Two years later, on 21 July 1974, the British ambassador to Ireland, Sir Christopher Ewart Biggs, was assassinated by an IRA car bomb. Provisional IRA brigades in England had further unleashed a wave of atrocities, notably the Guildford and Birmingham pub bombings in October and November 1974. The old neighbours were obliged to engage about the rancid political mound piling up in their shared back yard, even if their respective views on how to deal with it differed.

Throughout 1974 and 1975, the Republic's National Coalition government and its Minister for Foreign Affairs, Garret FitzGerald, grappled with the knowledge that not alone were the British persistently engaged in secret talks with the IRA, with whom his own administration would have no truck, but that the Prime Minister, Harold Wilson, was toying with the idea of a British withdrawal from Northern Ireland. FitzGerald deployed subtle diplomacy, without letting the British know that he was aware of their inclination towards withdrawal, in case it might encourage them.

FitzGerald quietly briefed key political journalists in Britain. At two informal dinners in west Cork in the summer of 1975 with Britain's Foreign Secretary, Jim Callaghan, FitzGerald warned him of the danger to Britain and north-west Europe of creating 'a situation in which extra-European powers such as the Soviet Union, China or Libya could meddle'.[10] Wilson's proposal for an independent Northern Ireland was eventually discussed by a cabinet sub-committee in December 1975. Opposed by Callaghan, the Chancellor for the Exchequer, Denis Healey and the Northern Ireland Secretary, Merlyn Rees, the proposal was scrapped, to the immense relief of the Irish government.

Recalling those events, Garret FitzGerald would later write: 'Looking back on that fraught period 30 years later, what remains most vivid in my mind about that time is the terrible sense of virtual impotence that I and others immediately involved felt in the face of the dangers which a British withdrawal would have created for our island, and our state. Neither then nor since has public opinion in Ireland realised how close to disaster our whole island came during the last two years of Harold Wilson's premiership.'[11]

* * *

Ireland also differed crucially from Britain in its enthusiasm for Europe. The attraction of Europe was mainly economic – and indeed bore fruit. By 1978 agricultural incomes doubled over 1970 levels as a result of Common Agricultural Policy subsidies to farmers, while urban Ireland benefited hugely from an influx of foreign direct investment, mainly US firms, in the 1970s seeking an English-speaking, low-cost, manufacturing base for their exports to the European market.[12] The subliminal mass appeal of union with Europe was the prospect of real economic independence through new markets for Irish produce, as an alternative to the status quo ante as 'the larder of England'.[13]

A 1972 referendum was carried by a landslide four to one majority and Ireland and Britain joined the EEC together in January 1973. Sean Lemass had overseen a new departure for the Irish economy and society in the early 1960s. By the end of the decade, the spirit of modernisation had taken root, and despite the troubles in Northern Ireland and occasional economic setbacks, the Republic, for the first time in its short history, was focused on the future, not the past.

Part of that future involved the development of infrastructure to meet the demands of expanding economic activity, including the supply of energy. Energy policy was largely dictated by the monopoly Electricity Supply Board, rather than the Department of Power, as it was then called. Labour Party TD Barry Desmond's quip that: 'From my experience of nearly ten years in the Dáil I have never regarded that Department, with no disrespect to [its] distinguished public servants ... as having a great reputation in terms of innovation and energy policy,'[14] would have resonated among his peers. From the spring of 1967, the ESB was actively researching the likely role of nuclear power in electricity generation, including its economics and organisation. The ESB advised the government that Ireland's first nuclear plant could be in operation by 1978.[15]

ESB engineers trained abroad, including at Harwell and Windscale, to build up the board's expertise and familiarity with nuclear reactors and technology. The search for a suitable location was under way and the board evaluated sites in the west and north-east regions, such as Easkey in Co. Sligo and Castlebellingham in Co. Louth, and in the south, including Carnsore Point, in Co. Wexford.[16]

Demand for electricity in the Republic was growing by an average 10 per cent a year by 1971, when the Minister for Transport and Power, Brian Lenihan, introduced legislation to create an independent regulatory agency, the Nuclear Energy Board (NEB), to advise the government on all aspects of the proposed new nuclear power plant. The political virtue of an independent advisory body was that it would fill the expertise gap on a subject about which politicians on all sides admitted they knew little or nothing. As a regulatory agency, the NEB would also act as a buffer, keeping the government of the day at one remove from regulation of the nuclear power programme.

Lenihan's Bill, and the prospect of nuclear power for Ireland, was met with universal acclaim in both the Senate and the Dáil. Indeed, Fine Gael spokesman, Tom O'Donnell, expressed disappointment that the government had not made a final decision in favour of a nuclear power plant.

'I have been making inquiries into the matter from people who are directly involved in the project and it appears that the green light was given to go ahead and to plan for a nuclear power station in 1978,' he contended.[17]

O'Donnell, among others, highlighted the importance of public information. Constituents attending his weekly clinics in Limerick were afraid of nuclear energy. Fear, O'Donnell suggested, 'engendered to a large degree by the outpourings of cranks, fanatics and pseudo-scientists and others, some of whom we have had in this country in recent times since the question of nuclear energy came to be debated ...'[18]

As Minister, Lenihan assured deputies and senators that any eventual project would have a North–South dimension in line with existing agreements on interconnection between the two states. The government – and the ESB's hope – was to purchase a 500 MW reactor, given that Ireland's total generating capacity at the time was 1,400 MW. By the time the nuclear power station was built, it was assumed Ireland's capacity would have reached 3,000 MW. Unless Northern Ireland took part of the output from the station, even a 500 MW reactor could prove difficult to accommodate without destabilising the national grid.

Lenihan estimated the capital cost of the nuclear power station at about £50m. For the industrially starved regions of rural Ireland, two thousand initial construction jobs and 500 permanent posts in the new power station looked enticing, as evidenced by the regular trail through the Department of local delegations, led by their public representatives, from areas lobbying for selection as the eventual project site.[19]

* * *

After sixteen continuous years in power, Fianna Fáil lost office to the National Coalition of Fine Gael and Labour under Liam Cosgrave in early 1973. The new government, elected on a platform of reform and innovation, had hardly had time to get their feet under the cabinet table when the oil crisis struck. Ireland's 65 per cent dependency on oil for its primary energy supply made the country especially vulnerable to the impact of the oil shock. On 23 November 1973, the government announced it had given the go-ahead to the ESB to construct a nuclear power station and to prepare a planning application for its preferred choice of site, Carnsore Point in Wexford.[20] A final decision on the location of the power plant would be made in 1975, the government said, and the estimated capital cost of the project was put at £100m.[21]

For the government and the ESB, the 1973 announcement represented the high point of approbation for a nuclear project in Ireland. Like BNFL, Ireland's electricity board was destined to discover that its nuclear plans, too, were far from secure.

12

Windscale to Sellafield

> 'Windscale did not become any less noxious or objectionable when it changed its name to Sellafield and continued its activities as before.'
>
> Irish Minister, John O'Donoghue, 2002

Depending on the source, the Director of the Windscale site in the mid-1970s, Peter Mummery, was either loyal and committed to his industry almost to a fault, or the archetypal Windscale man, one of the 'old dinosaurs', arrogantly dismissive of any challenges or criticism of the nuclear industry, who would take on all-comers internal or external on that account. Windscale was advancing into the next phase of its existence; from a military site with an add-on trade in civil nuclear fuel services to an internationally trading commercial business operation. But the transition was not without its difficulties.

The military aspect of BNFL's business, intrinsic to its finances, operations and culture, was barely acknowledged by the company.[1] BNFL's tendency to sublimate Windscale's military inheritance and its furtive continuing involvement in making bomb materials contributed to anti-nuclear campaigners' general disdain for the company and their designation of all its activities as infernally inspired.

The corporate livery and the logo on the company notepaper might have changed, but the formation of BNFL had involved more a wholesale transfer of personnel from the UKAEA, with an organisational culture already thirty years in the making, than any dramatic influx of new blood or ideas. BNFL and the UKAEA even shared the same chairman, Sir John Hill.

The public and political environment into which the new BNFL was thrust was also different. The company and its personnel struggled to adapt to the requirement for a new style of open and effective communications, antipathetic to its traditional corporate culture which had been honed as a guardian of official secrets and pioneers in a vital national undertaking. At Windscale, this culture had long since mutated into a siege mentality that bordered on paranoia, symptoms of which were already apparent by the time of the Windscale fire.

In 1975, BNFL hired Harold Bolter, a journalist with the *Financial Times*, to head up its corporate communications. Bolter responded to the 'nuclear dustbin' tag by organising media visits and public meetings at BNFL's shipping terminal in Barrow-in-Furness, by now a regular protest venue for environmental groups campaigning against Japanese shipments. Next came a high-profile public debate at Church House in London, in January 1976, at which Tony Benn was the opening speaker. BNFL's campaign to gain public acceptance for its overseas reprocessing made enough progress by March that year to persuade the cabinet and the Secretary of State to grant formal approval for the overseas contracts plan.[2] Finally Bolter, on his own account and against the gut instincts of the Site Director, Peter Mummery, who had drafted in constabulary from every nuclear site in Britain to cope with anticipated breaches of security, turned a planned day of mass protest at Windscale by Friends of the Earth supporters in July 1976 into a day of dialogue with BNFL on the periphery of the site.[3]

Effectively, BNFL had seen off the initial media challenge by Friends of the Earth. The company began pointing to the dangers of any further public controversy over its plans. Negative publicity caused uncertainty and misgivings amongst its prospective customers, particularly in Japan, BNFL argued. In February 1976, as if to confirm its claim that undue delays in granting permission for Thorp would jeopardise lucrative foreign business, BNFL announced that its partner in United Reprocessors, Cogema, would take a half share in the Japanese contracts. The French company also set about negotiating contracts for European business, of which, in the end, it secured the lion's share. As represented by BNFL, dividing the Japanese contracts with Cogema was a business loss occasioned by anti-Thorp campaigners and lack of resolute backing for the industry by the government of the day. Other sources within the industry suggest that the Japanese were nervous of placing all their spent fuel eggs in one reprocessing basket and the carve-up of the business occurred by mutual agreement between BNFL and Cogema.

BNFL's plans were also vulnerable to the UK planning process, specifically negotiating the first hurdle at local authority level. Members of the planning committee of Cumbria County Council came under relentless pressure from all sides as, under the glare of national publicity, they considered the grant of planning permission for Thorp. Friends of the Earth derided the council's credentials. In their view, Cumbria County Council was 'usually concerned with applications for new garden sheds and the like'.[4] By now Friends of the Earth and their allies had switched the focus of their campaign to a demand for a full public enquiry into Thorp. The Secretary of State for Environment must 'call in' BNFL's planning application. At a public meeting in Whitehaven in late September, Peter Mummery warned the council that any delay in granting planning permission 'despite the government's support of this business'[5] would threaten overseas contracts and 5,000 promised new jobs at Windscale.

The council refused to be browbeaten. Instead, its planning committee delivered a fudge. On 2 November 1976, Cumbria County Council announced that while they were minded to approve BNFL's application, it constituted a departure from the county development plan. The ball was passed back to a reluctant government.

BNFL was confident the cabinet would find in its favour and direct the county council to approve planning permission for Thorp, though they were discomfited when as late as 25 November, the Environment Secretary, Peter Shore, told the Commons he still had not made up his mind. Within a few days, their cause was lost, not because of lobbying by environmentalists or murmured threats of legal action if the government approved the Thorp plan, but because of a leak of tens of thousands of litres of contaminated water from a silo on the Windscale site.

* * *

Engineers excavating for an extension to the existing silo uncovered the leak in early October 1976. The silo, B38, built in 1964, had a capacity of 70,000 gallons, and was used to store the cladding stripped from magnox fuel elements prior to reprocessing. The contaminated cladding was stored under water.[6] It was this radioactively contaminated water that was seeping into the ground at a rate of 400 litres per day. There was no certainty as to how long this had been going on, and no one had any idea of how to stop it. But since it was initially judged to have no implications, either for the local water table or as an underground plume heading towards the sea, it was simply reported as a matter of course to the regulator, the Nuclear Installations Inspectorate, on 22 October, and five days later to the Department of the Environment.

According to Harold Bolter, 'nobody bothered to tell anyone at the company's headquarters'. The first information he personally received came from the local MP for Copeland, Jack Cunningham, who had picked up a rumour circulating in the House of Commons.[7] Secretary of State Tony Benn was not advised until 8 December. His reaction was predictable. The Windscale management's excuse that the leak had no significant safety or health implications, and publicising its discovery might rock the boat at a time when the local council was considering the planning application for Thorp, only put a further dampener on the Energy Secretary's Christmas spirit.

Benn accused the company of a deliberate cover-up and decreed that in future, all incidents at the site, even the most trivial, must be immediately reported to him. At cabinet, a furious row erupted over the Thorp plan and on 22 December 1976, Peter Shore announced a public enquiry into Thorp.

The Windscale Inquiry, under Mr Justice Roger Parker, opened in Whitehaven's Civic Hall on 14 May 1977 and sat for a hundred days. In the course of the inquiry, Friends of the Earth and their allies sought to place Britain's entire nuclear programme at Windscale on trial. BNFL based its case for Thorp on economics; the anticipated continued expansion of nuclear power in the UK and the financial benefits to Britain of its foreign contracts. Economic data won out over the environmentalists' exposition of assumed risks attendant on the Thorp project and details of the site's safety performance to date.

The inquiry report, finally published in March 1978, left Friends of the Earth and their supporters bitterly disillusioned. 'From that day onwards,' Walt Patterson wrote 'even those critics who had looked for dialogue and rational discussion decided that official nuclear policy was beyond influence by rational argument.'[8] A year after the Windscale Inquiry had opened, MPs finally gave the go-ahead for Thorp in a House of Commons vote, with 224 votes in favour to 80 against.

BNFL had apparently won the first major battle with environmental groups in a war of attrition that would continue for the next twenty years. Even this initial victory came at a price, part of which was the loss of its relative media anonymity: what might previously be concealed on a military site as a matter of national security was now openly subject to judgement in the court of public opinion. Every incident at its Windscale site, however minor, could be amplified by its critics way beyond any proportionate risk to public health and safety. Not that Windscale was averse to fashioning sticks with which to beat itself.

In April 1977, general site safety was called into question when an essential shipment of nitrogen to the site was blocked by workforce pickets. A strike by the Windscale workforce that had already lasted for a month arose from a pay claim by thirty-two changing room attendants. When the attendants withdrew their labour, three thousand general workers were sent home on full pay. The Windscale workers subsequently refused to use changing rooms manned by members of the management team. In retaliation, the management stopped their pay and the dispute escalated into a general strike. Much to BNFL's resentment, Tony Benn personally intervened to force management concessions to bring an end to the strike.[9] Shortly afterwards, Peter Mummery was moved from his post as Windscale Site Director to become BNFL's Director of Health and Safety at its Risley headquarters.

Boreholes drilled to generally assess levels of site contamination in March 1979 revealed a further substantial radioactive leak into the ground, this time of highly active liquid waste, or HAL. The leak had been going on for at least eight years. The cause was a faulty valve leading to a holding tank in a long abandoned facility, B701, into which the HAL was pumped to provide sample material for research at Harwell in the late 1950s into waste management technology.[10] By the time the source of the leak was located and the valve sealed, a month after its discovery, it was estimated that 10,000 litres of the highly active liquid, possibly containing between 30,000 to 100,000 curies of radioactivity, had been escaping into the ground since the late 1960s.

Nor was anyone convinced that these two incidents represented the total extent of ground contamination at the site.

'It was often said when I worked for BNFL that if you dug a hole virtually anywhere on the Sellafield site there was every chance that you would find some radioactive contamination,' Harold Bolter would later write.[11]

Two months later a fire in B30, the head end plant for magnox fuel, led to a shutdown of the plant for several weeks, further increasing the backlog of magnox fuel in Windscale's cooling ponds and exacerbating its progressive deterioration. By this stage, BNFL's regulator, the Nuclear Installations Inspectorate (NII), had already decided to conduct a review of safety at Windscale. Now it was directed to do so by the new Conservative government under Margaret Thatcher that had replaced Jim Callaghan's embattled administration in 1979. The NII report, when it was finally published in February 1981, made fifteen detailed recommendations for safety improvements at the site.

'By the early 1970s the standards of the plant at Windscale had deteriorated to an unsatisfactory level. We consider this represented a poor base line from which to develop high standards of safety. We are strongly of the opinion that such a situation should not have been allowed to develop, nor should it be permitted to occur again,' the NII concluded.[12]

* * *

BNFL embarked on reorganisation of the Windscale management structure, combining Windscale Works with Calder Hall, and, in a move decried as the ultimate PR stunt by its critics, changed the name of the site back to Sellafield, as it had always been locally known in West Cumbria. What they couldn't change were the systemic flaws that lay at the root of the safety problem, including what Christopher Hinton had sought to point out twenty years earlier when he voiced concerns about high-level waste storage to his minister and officials[13].

Complacency and management incompetence were hardly unique to Britain's nuclear industry in the 1970s, but at every level of the British establishment there was a deafening silence about the lack of investment over several decades in programmes for waste management control systems, from discharges to the Irish Sea to the design, construction and operation of actual facilities on the site.

Reprocessing, whether for military or civil purposes, was an inherently expensive enterprise. The real economic cost of producing military plutonium and, later, of dealing with the spent fuel from the civil reactor programme included waste management technologies to minimise discharges to the environment and avoid the stockpiling of a legacy of hazardous waste materials on the site, or at least contain them effectively. Contracts with the Ministry of Defence or with the electricity supply boards should, by rights, have reflected these wider costs. They didn't.

At Windscale, reprocessing was being done on the cheap as a matter of public policy. The series of high-profile incidents of the late 1970s and early 1980s were, at least in part, a direct consequence of this policy.

Windscale was also a uniquely complex industrial site, in which a range of different processes interacted with one another and at each stage involved the management and control of hazardous radioactive materials whose behaviour was unpredictable and often unforeseeable. Further, a technical hitch in one area might have a domino effect throughout every stage of the processing of spent fuel, creating new unanticipated hazards further down the line. There was less room for human error than in comparable industrial processes without a radiological component. The addition of radioactive materials to normal chemical processes magnified the potential for catastrophic consequences of even the most mundane accident.

The increased fission content of high-level waste from reprocessing of civil magnox fuel in the early 1970s usefully illustrates this point.[14] The higher volume of high-level waste it created put pressure on existing tank storage space and ultimately forced a shutdown of magnox reprocessing for several months in 1972 until new tanks were built and installed. This, in turn, gave rise to prolonged storage of magnox elements in the Windscale ponds. Fuel element decay contaminated the pond water which, without regular purging to sea, risked unacceptably high levels of exposure to contamination of site workers.

Avoiding contamination on the site itself justified the Windscale management seeking authorisations for ever-increasing radioactive discharges to the Irish Sea. But while Windscale discharges of caesium and plutonium to the Irish Sea in the mid-1970s may have been within authorised legal limits, they were reaching a level that was clearly unsustainable. Many years later a University College Dublin research team would note: 'Past levels of liquid radioactive discharges from Sellafield, most notably those in the mid- to late-1970s, were a matter for serious concern as committed effective doses to the most exposed members of the UK public reached 50% of the then annual limit of 5 mSv recommended by the International Commission on Radiological Protection (ICRP).'[15]

* * *

Accepting the radiation risks was one thing when the defence of your country was at stake. With the expansion of the civil nuclear industry, the ordinary man or woman in the street was entitled to take a more subjective view of the hazards to which they themselves, their community and the general environment were exposed. By the mid-1970s, nuclear safety had become a central political issue throughout the western world and any example of what, only in theory, might result in an environmental catastrophe did little to reinforce public confidence in the industry.

The dissident Russian biologist Zhores Medvedev was exiled from the Soviet Union in 1973. Writing in the *New Scientist* in November 1976, Medvedev cited a nuclear accident that, he said, had taken place in the southern Urals some twenty years earlier and resulted in widespread devastation. Reports in 1958 that a mysterious nuclear accident had occurred in Russia had been rejected by the then chairman of the USAEC, Lewis Strauss. There was 'no intelligence' to support the view that any such incident had occurred, he said. In the intervening twenty years, the story was forgotten.

Medvedev's article became a global news story overnight, as did its trenchant dismissal by BNFL's chairman, Sir John Hill, as 'rubbish' and 'pure science fiction'.[16]

The Mayak complex near Kyshtym in the Chelyabinsk province in the southern Urals was the Soviet equivalent of Windscale. Mayak was a site, however, without any safety system worthy of the name and less in the way of emergency plans to protect the workforce, the local environment or the local population from the consequences of radioactive pollution. Highly active liquid waste from its plutonium production facility was pumped directly into the local Techa river.

In 1953, the residents of Metlino village, about seven kilometres downstream from the point of discharge, were evacuated as their environment progressively became ever more hopelessly contaminated. Between 1955 and 1960 the inhabitants of a further nineteen settlements along a 100 kilometre stretch of the banks of the Techa, about 7,500 people, were also moved away from the river. A number of crude open-air 300 cubic metre tanks were constructed on site to take the high-level waste. Material was also diverted into a nearby lake, Lake Karachoy.

On 29 September 1957, about three weeks before the Windscale fire, one of the tanks exploded with a force equivalent to about 100 tonnes of TNT. About 90 per cent of the radioactive fallout deposited locally but a plume of radiation, mainly caesium and strontium, rose one kilometre into the atmosphere and contaminated an area of between 1,500 and 2,300 square kilometres with a resident population of about 270,000 people. Altogether, 10,730 local people were evacuated from the immediate area.[17] Ten years later, in 1967, the heat from the radioactive waste dried out the edges of Lake Karachoy. Intensely radioactive materials were dispersed on the wind, polluting up to 20 square kilometres of the surrounding countryside.

Sir John Hill's sensitivity to the suggestion that any such waste tank could explode was understandable, given that Windscale by now had several tanks filled with highly active liquid wastes and the expansion of reprocessing would ultimately give rise to a requirement for twenty-one tanks, from the original six built in the 1950s.

Windscale's tanks were designed around the principle of 'defence in depth', with several layers of protective mechanisms, from internal cooling coils to independent water supplies, to spare tank capacity, all intended to ensure the highly active liquid could not overheat, and if it did, or if there was any other risk of tank failure, to enable prompt intervention to negate any possibility of an explosion or major release of radioactivity. Public attention to the possibility that a high-level waste tank could explode was particularly unwelcome when BNFL's plans for Thorp were the centre of an acrimonious public wrangle and public confidence in nuclear expert opinion was rapidly beginning to dissipate.

13

An Acceptable Risk

'We understand and recognise that the ESB and the Nuclear Energy Board are not maniacs likely to poison us all in a fit of absent-mindedness.'

Fine Gael Deputy John Kelly (1971)

By the mid-1970s, the focus of anti-nuclear protests had long since moved away from local concerns about the immediate environmental impact of nuclear power stations – such as the ecological impact of discharges of heated water on local waterways or the effect of steam emissions from cooling towers on regional meteorological conditions – to fear of a catastrophic accident and the consequences for public health and the environment of the widespread dispersal of radioactive contamination.

Anti-nuclear campaigners, especially Greenpeace, forced the nuclear option to the political centre stage through a combination of demands for public inquiries and plebiscites, high-profile PR stunts, mass protests and court challenges to the planning system.

Even for the most ardent anti-nuclear campaigners, it would have begged incredulity to claim a nuclear power station could explode with anything like the force of a nuclear bomb. In any case, conventional wisdom amongst engineers and scientists at the time was that a nuclear reactor could not explode. Studies on the bomb survivors of Hiroshima and Nagasaki provided the benchmark against which the health effects of radiation poisoning from a reactor accident could be gauged as well as a reminder of its deadly consequences.

The case of Japanese fishermen contaminated by fallout from the US Bikini Atoll H-bomb test in 1954 had provoked an international outcry.[1] The US administration was exposed by Joseph Rotblat, research professor with St Bartholomew's Hospital in London, as having minimised the fallout figures, although Rotblat himself later conceded his own calculations exaggerated the extent of the fallout and, therefore, its effect. The seeds of public concern over health risks from radioactive fallout were well-sown by this and other military mishaps of the 1950s and 1960s, which anti-nuclear activists could now turn to effective use in their propaganda.

The anti-nuclear movement also developed the tactic of meeting like with like: if the nuclear industry positioned itself as the sole repository of science and wisdom on nuclear technology then the NGOs would develop their own base of expertise.

'We, as Greenpeace, did the political angle, writing the policy,' John Bowler, Campaigns Director of Greenpeace Ireland for ten years from 1987, later explained. 'But we had scientists working on our documents – John Large, who's probably the best known; Frank Barnaby, and other people – whose credibility can't be questioned in terms of their knowledge of the issues and where they've come from and who themselves had worked for governments.'

Caught between the claims of the industry experts and those of the anti-nuclear NGOs, the public was entitled to feel confused: scientists on both sides argued their assessments and findings were conclusive, usually in opposite directions. In reality, the most they could do was define any risk and issue a probabilistic estimate of its consequences, an uncertain process inevitably coloured by the scientists' own political and organisational perspectives, and hardly satisfactory to a public and media seeking certainty.

The publication in 1974 of the Rasmussen Report by the Massachusetts Institute of Technology, following a three-year $4m study of reactor safety, delivered partial public reassurance that nuclear power was safe. Early press reports highlighted a conclusion in Rasmussen's executive summary: the chances of a disastrous nuclear accident were about one in a million years. The real significance of Rasmussen to nuclear engineers and scientists, however, was that, for the first time, it acknowledged the possibility of a core meltdown accident, a scenario the industry had previously rejected as 'incredible'. But, Rasmussen suggested, this was not necessarily the disaster portrayed by the anti-nuclear lobby. Only in a very small fraction of cases would core meltdown be synonymous with a massive environmental release of radiation. A reasonable core containment system might hold for several hours, allowing time for human intervention to ameliorate the danger. Nuclear power was an acceptable risk, the industry experts concluded.

* * *

'The whole fission power process is "unforgiving", so far as human errors are concerned,' Christopher O'Farrelly, a member of the ESB nuclear project team wrote in August 1971, in a report to the board on 'Nuclear Reactor Safety'. O'Farrelly calculated the likelihood of a major nuclear accident as one in every 10,000 reactor years, or about every thirty years. He predicted, more presciently than he could have known at the time, that 'there will be at least one major reactor disaster in the world during the life of the ESB's first nuclear station'.[2]

Nuclear engineers, gathered at a convention in Schenectady, New York State, in 1966 had been addressed by Edward Teller on the subject of reactor safety. Teller was dismissive of existing reactor containment systems, in his view likely to be as effective in containing the shards of a disintegrating reactor as an eggshell. Teller believed a major civil nuclear accident was inevitable sooner or later and urged the industry to put reactors underground, ideally under thirty metres of clay or rock. Placing reactors underground would provide dual protection: preventing a breach of containment if the reactor exploded as well as obviating the risk of a 'China syndrome'[3] type meltdown.

Impressed by Teller's arguments, O'Farrelly strongly recommended siting the ESB's first nuclear power station underground. Although it would add about £5m to the eventual cost of the project, it was the only way in which a truly 'independent' containment system could be achieved, he contended.[4]

'The probability may be extremely small, but unless the reactor is underground, the potential for a disastrous accident will always be present,' O'Farrelly wrote.[5] 'It does not seem justifiable to skimp on safety provisions in order to give nuclear power a marginal economic viability.'[6]

An underground reactor had distinct PR advantages. 'With overground siting, public relations will always be tricky,' O'Farrelly observed. 'This is quite inevitable, as [the ESB] has to perform a balancing act between equivocation and the complete honesty which it would probably be folly to indulge in.'[7]

The ESB was at the mercy of the market in its search for a reactor of proven design that would prove suitable for Irish conditions. Britain's gas-cooled model, the AGR, of which several of the engineers in the ESB nuclear group had direct experience, was rejected early on.

'It was fairly obvious that their nuclear programme of gas-cooled reactors was not the way to go. It was not economic,' Sean Coakley, another member of the project team, noted many years later.

Instead, the ESB team looked to American designs, an LWR or PWR similar to those operating throughout the US and Europe or the Canadian heavy water reactor design, the CANDU. Very soon they realised their choices were limited to whatever the manufacturer was prepared to supply and this didn't include building the station underground. A location remote from large population centres with an offshore prevailing wind was the next best available option to minimise the environmental impact of any nuclear accident. From this perspective, ESB engineers concluded Carnsore Point 'was probably the best site for a nuclear power station in the whole of Europe'.[8]

* * *

By any standards, the ESB's research into every aspect of its nuclear project, including environmental impact, health and safety, reactor selection, local ecology and archaeological data on the Carnsore area, was both thorough and impressive. More than one hundred reports and independently commissioned studies were produced between 1967 and 1980. By the time the ESB had finalised its draft specification for the project and submitted it to external organisations for comment in 1975, what had started out as the search for a standard plant of proven capability had accumulated such stringent safety features and design requirements that it was considered quite draconian.[9]

Ireland's nuclear energy plan was temporarily put on hold in 1975 when electricity demand plummeted from 10 per cent growth per annum to zero in the Republic, which like much of the rest of Europe was struggling with the economic recession caused by the first oil crisis.[10] The new national regulator, the Nuclear Energy Board (NEB), established by statute in 1971, had finally sprung to administrative life in 1975 with the appointment of a Chief Executive, Chris Cunningham.

A native of Donegal, Cunningham had spent several years in the US working on reactor systems. At the time of his appointment to the NEB, Cunningham was an assistant Chief Engineer with British Nuclear Design and Construction Limited, where he was responsible for nuclear fuel element development and nuclear science services. Cunningham had worked on the design of fuel elements for Britain's early magnox reactors.

The NEB received the ESB's plans for the Carnsore site from the Department of Transport and Power in November 1975 and promised to report back within six months.[11] The ESB proposal was assessed with the aid of experts from the European Commission. In its report to the department in mid-1976, the NEB noted: 'The site at Carnsore compares very favourably with sites selected for similar nuclear power stations in North America and Europe.'[12] The NEB would require more detailed submissions from the ESB before it delivered a final verdict, but its interim conclusion was that the Carnsore project presented 'no major safety problems'.

* * *

This was a view no longer universally shared. As politicians' antennae picked up increased levels of anxiety among their constituents and party supporters about a nuclear Ireland, the previous political consensus on the desirability of an Irish nuclear project began to unravel. The battle for the hearts and minds of the Irish public would be fought out between the ESB and the government on one side, since the NEB declared their neutrality from the outset, and a form of popular protest that was new to Ireland.

A motley coalition of far-left political groups, uneasy bedfellows with politicians from the more mainstream political parties keen to keep in step with the popular mood; a new breed of environmental activists; the Irish feminist movement; popular musicians such as the balladeer Christy Moore, Clannad folk group and Chris De Burgh; John Carroll, leader of the country's largest trade union, the Irish Transport and General Workers' Union and the international campaigner, Petra Kelly, who at the end of the 1970s became the co-founder of Germany's Green Party, all united around the banner of 'Nuclear Power – No thanks' to oppose the Carnsore project. This colourful, multi-faceted protest movement, with its street pageants depicting nuclear horrors, folk music concerts interspersed with anti-nuclear sketches, voluminous pamphlets and anti-nuclear stickers and posters and mass rallies at the Carnsore site itself between 1978 and 1981, attracted media coverage way in excess of its actual strength or levels of popular support.

The first Carnsore concert in August 1978, a highly organised event despite a torrential downpour that failed to dampen the protestors' enthusiasm, was attended by at least 5,000 people, bussed in from around the country.

The anti-nuclear movement in the main attracted students and young people, the first generation of Irish people who had grown up with any confidence or conviction of having a real stake in their country's future. On the Sunday of the weekend protest, following mass in the local church, a small group of local people came to the proposed reactor site where the protest rally and concert had been staged. The visiting protestors were by this time engaged in building a cairn of stones on the beach at Carne, as a monument to their opposition to the nuclear project. The locals said the rosary, but it was unclear if they were praying for deliverance from nuclear power or from the hordes of young people who had invaded their small community over the weekend.

One group, the Cork Anti-Nuclear Alliance (CANA), would shortly discover a genuine local target: a nuclear student training reactor in their own backyard. Frank Aiken's 'magnificent gift' from the US Atoms for Peace programme had long since been installed in the science department of University College Cork.[13] Calling it a reactor was a bit of a misnomer, according to Dr William Reville who was UCC Radiation Protection Officer at the time, since the device 'bore as much resemblance to a normal reactor as a candle to the sun'.

The student training reactor consisted of a five-foot high, 24-inch diameter, stainless steel water-filled tank, in which several hundred uranium fuel slugs were arranged in an aluminium lattice around a neutron source at the centre, containing a small quantity of plutonium and beryllium. The sub-critical device – it was engineered to ensure it could never go 'critical', producing a sustained chain reaction – was used to demonstrate simple experiments in nuclear physics to students. It was installed in the early 1970s and retrospectively licensed by the NEB. No one in the college or the physics department thought much about it, and hardly anyone in the broader community was aware of it, until a national radio programme advertised its existence to early-morning listeners in hyperbolic terms, just as the Carnsore protests were reaching their national crescendo.

William Reville claims this was the only period in his entire life when he actually lost sleep over a work-related issue. CANA campaigned to have UCC's licence to operate the device withdrawn. For a while, media interest was intense, a matter of some enjoyment to the Professor of the Physics Department, Frank Fahy, who was happy to point out to reporters that the greatest danger from the heavy uranium slugs might be to drop one and break your foot. Dr Reville, meanwhile, coped with the almost daily four-page letters from CANA, to which he felt obliged to reply in corresponding detail and worried that the NEB might indeed revoke UCC's licence.

CANA variously accused the college of cover-ups and secrecy about the reactor and of breaching its licence conditions, but Dr Reville believes they ultimately overreached themselves in the ferocity of their accusations and insinuation of a 'capitalist plot', damaging their credibility with the media.

In the Dáil, the Labour Party deputy, Ruairi Quinn, who was later keen to have all of Ireland declared a nuclear weapons and civil energy free zone,[14] championed their cause. Quinn questioned the then Minister for Energy, George Colley, about 'evacuation plans' for Cork city and its environs and suggested the UCC reactor might compromise the proposed public inquiry into the Carnsore project.[15]

Cork City Council became the first local authority in the Republic, in 1981, to declare itself a 'nuclear free zone'.[16] The council motion proposing this new status carried an addendum making a special exemption for the UCC reactor. Dr Reville wrote to the local authority requesting to know what was meant by this exemption. A few months later he received a reply from the local authority chief engineer, who advised he might better direct his enquiries to the councillors who had proposed the motion in the first place. A nuclear free zone sign was erected on the main road from Dublin entering Cork city, where it stands to the present day.

The student training reactor was dismantled in the early 1980s to make space for new lecture theatres. The steel tank was broken up and sold for scrap. The 5,500 lbs of natural uranium and the plutonium/beryllium source were placed in storage and maintained under licence at the university, subject to IAEA safeguards. Various efforts were made in the intervening years by the college and the Radiological Protection Institute of Ireland (RPII) to have the fuel and the neutron source repatriated to the United States, an operation that apart from generating a new media storm would likely cost in the region of several million euros. As of 2005, the Americans still politely pointed out that the training reactor was a gift and, as such, they couldn't take it back. As an RPII officer put it, Aiken's 'magnificent gift' had turned out to be 'a most unfortunate gift' indeed.

It has to be said that if all those in Irish public life who later claimed to have discovered their anti-nuclear credentials at the Carnsore festivals had actually been there, that corner of Wexford should have broken off into the Irish Sea under their cumulative weight. Yet for all the saturation media coverage, the anti-nuclear protest movement was not the decisive 'people power' factor in the ultimate demise of the ESB's plans some later claimed it to be.[17] Events elsewhere, and economics, proved far more crucial in determining the fate of Ireland's nuclear project.

14

The Legacy of Three Mile Island

'Do we want to live in a land laid waste by a nuclear accident? Do we want our children to grow up with horrific injuries and other medical problems? They may not grow up at all if they are exposed to a nuclear accident.'

Senator Margaret Cox, 1998

At 4 a.m. on March 28 1979, pumps feeding water to the steam generator at the Pressurized Water Reactor No. 2, TMI-2, at Three Mile Island nuclear power plant near Middletown, Pennsylvania, malfunctioned.[1] The steam generator and the reactor itself shut down, but residual radiation levels in the core caused its water coolant to heat and expand. With no steam generator to draw off excess heat from the coolant, pressure in the reactor increased. A safety valve opened to relieve the pressure. The valve should have shut within ten seconds of the core pressure returning to normal. The plant operators, misled by confusing indicators on the panels in the control room, assumed that it had.

Instead, the valve remained open for two and a half hours, allowing tens of thousands of gallons of coolant to drain away, leaving the upper part of the reactor core uncovered. At least one-third of the fuel in the core melted under its own heat. Attempts by the operators to stabilise the reactor only compounded the damage as the core was successively covered and uncovered by coolant over a period of several hours. Zirconium cladding on the reactor's fuel rods reacted with steam, producing a hydrogen bubble at the top of the exposed core.

For the next five days, the plant operators, the Metropolitan Edison Company (Met Ed), state and federal agencies and officials, including the office of the President of the United States, Jimmy Carter, struggled to assess the extent of the damage. Widespread confusion reigned and contradictory information was relayed about how much radiation was being released to the environment, and whether evacuation of local residents was necessary. Concern about the risk of a hydrogen explosion that could rupture reactor containment and cause extensive contamination peaked on 30 March, but lessened as the condition of the reactor core stabilised in the following days.

About two weeks before the accident, on 12 March 1979, a Hollywood thriller, *The China Syndrome*, depicting a near meltdown in a nuclear reactor, had opened in New York. The attitudes of media descending on Middletown, a town of about 25,000 people, were arguably more informed by the Hollywood blockbuster than official statements from Met Ed, the State Governor's office or the Nuclear Regulatory Commission. Reporting ranged from cheap sensationalism, such as the TV crew who asked local residents to remain off the streets of Middletown as they filmed in order to give the impression of a ghost town, to misleading headlines slavishly repeating whatever official line was on offer.[2] Since the official lines, at best, were either overly technical or hopelessly confused, they served more to increase terror amongst local people than assuage it.

On 30 March the State Governor, Richard L. Thornburn, proposed the evacuation of pregnant women and young children within a five mile radius of the plant. Other residents were advised to shelter in doors and keep their windows closed. About 100,000 people temporarily left the area.

The accident itself was the result of technical failures compounded by human error. However, the communications effort by Met Ed, the Nuclear Regulatory Commission and most of the political actors who took centre stage at various points in the unfolding drama turned it into a complete PR disaster. Not only was TMI-2 destroyed; so too was the confidence of the American public in the safety of nuclear power for the next quarter-century.[3]

The 900 MW Babcock and Wilcox-designed reactor had only been commissioned into service a mere three months earlier. TMI-1, its sister unit, had been operating since the mid-1970s but was closed down for maintenance at the time of the accident. TMI-1 remained shut for several years, reopening only in 1985. Met Ed was brought to the brink of bankruptcy, but not just by the loss of revenue in electricity sales from its crippled asset. Clean-up costs for TMI-2 were estimated near $1bn, about twice what the unit had cost in the first place.

A major international news story, the Harrisburg accident as it became known,[4] caught the imagination of anti-nuclear activists across Europe and reignited public fears about the safety of nuclear power. In Germany, the governor of Lower Saxony announced that planning permission would be refused for the reprocessing plant at Gorleben, not on safety grounds, but simply because it was no longer publicly acceptable. Plans for a reprocessing plant at Wackersdorf in Bavaria were also shelved. The German authorities decreed all spent fuel produced in German nuclear power plants would be shipped to Sellafield or the Cogema plant at La Hague for reprocessing.

In France, Three Mile Island sparked media calls for a national public inquiry, but the French government kept faith with its nuclear energy policy. There were mass demonstrations in Sweden, where the decision to go nuclear had long since been mired in political controversy, and had already driven one government from office. Within a year of the TMI accident, Swedish voters only narrowly passed a referendum that favoured continuing the country's programme, but cut the projected number of stations from twelve to six and promised a phase-out of nuclear fission power within twenty-five years. Austria, which had already constructed a new nuclear power station at Zwentendorf, near Vienna, put the project on ice. Prime Minister James Callaghan based his message of public reassurance on the fact that Britain had opted for its own unique advanced gas-cooled reactor design. 'I can assure the country that the incident which took place in Harrisburg could not take place here,' he said. 'We have been very wise in concentrating on a safer type of reactor.'[5]

* * *

The events at Three Mile Island sounded the death knell for the ESB project. It was a matter of economics, though reality took a while to sink in.

'The accident at Three Mile Island had a disastrous effect on the nuclear plans of all small electric utilities for, not only did it provide a bonanza for the anti-nuclear lobby, but it also made a deep impression on utility managements,' Christopher O'Farrelly would later suggest. 'The possibility of a major power station being written-off after a few weeks service was a veritable nightmare for any financial director and there was little consolation in the knowledge that clean up might cost as much as the original plant.'[6]

The Fianna Fáil government that had swept to power in 1977 with the largest parliamentary majority ever in the history of the state promised full employment by reducing the numbers of registered unemployed by a targeted 25,000 per annum and generating an annual 6.5 per cent increase in economic growth. A new Department of Economic Planning and Development was established to seek out the holy grail of prosperity in circumstances that from a national perspective might have appeared propitious, but would shortly collide with some uncomfortable international realities. The party's election platform had included other measures of undoubted populist appeal, including the elimination of private housing rates, previously the main source of funding for local authorities, and private motor vehicle taxes, measures that would significantly dent the availability of state funds for investment in basic infrastructure should the anticipated boom fail to materialise.

The Fianna Fáil administration also inherited the ESB nuclear project from the National Coalition. In March 1977, a few months before the election, the Fine Gael Minister, Peter Barry, had challenged the anti-nuclear movement and what he perceived as an ambivalent Fianna Fáil attitude towards Ireland's nuclear project.

'At the moment the only alternative to [fossil fuels] seems to be nuclear energy and we will have to make way for that,' Barry said. It was a test of democracy.

'It does not matter whether it is dear or cheap energy, oil or gas or turf, there is the possibility of no energy, if the people in democracies who are now objecting to the establishment of nuclear energy stations get their way and these stations are not established,' he stated.

In terms reminiscent of Ernest Bevin's post-war rationale for Britain's atom bomb project, Barry argued that if Western governments, including the Irish government, failed to risk difficult and unpopular decisions on nuclear energy, then: 'the great threat of domination by the east will not alone happen but it will not be a battle because they will have nothing to fight to take over'.[7]

* * *

The new Fianna Fáil minister with responsibility for energy, Des O'Malley, soon came to know just how difficult and unpopular the nuclear portfolio might prove to be. Pilloried and personally vilified by the anti-nuclear movement and criticised by his political opponents for retaliating when he dubbed his tormentors 'flat-earthers',[8] O'Malley stood his ground against opposition proposals for a joint parliamentary committee to assess the Carnsore project. For over a year, he also held out against increasingly vociferous demands for a public inquiry into the project, from both the opposition and the media, most notably, Fianna Fáil's own paper, the *Irish Press*, on the basis that it would serve no useful purpose.

In the short term, Ireland faced a crippling energy crisis. Electricity demand in the Republic had increased by 10.5 per cent in the year to March 1978 and there was an expectation of an average 8 per cent annual rise throughout the next decade. More important, Ireland was about 75 per cent dependant on oil for electricity generation. Even with the diversion of 50 per cent of the output of the recently discovered Kinsale gas field to the ESB for generation purposes, dependency on oil would continue to increase.

Internationally, the problem was not so much the price of a barrel of oil; it was one of supply. The onset of the second oil crisis, borne out of political turmoil in Iran that led to ousting the Shah in February 1979 and his ultimate replacement by the Islamist, anti-Western regime of the Ayatollahs, had once again pushed energy to the top of the agenda for most European governments. The second oil crisis took a particular toll on an Irish economy that was ill-prepared to adjust to it.

'Because of the shortage of oil, because of our very high dependence on it and because of our inability to switch quickly to other things we were, needless to say, in a very difficult situation,' O'Malley later observed.

Brown-outs and blackouts were the order of the day. The brown-outs were particularly serious for industrialists as sudden drops in the level of power risked irreparable damage to sensitive machinery in factories. Holding the dual portfolio of industry and commerce with energy, O'Malley was made sharply aware of industry complaints. Nor could he ignore the anger of ordinary consumers about the constant breaks in electricity supply. The level of voltage was also by no means uniform throughout the country and areas of low voltage, especially in the west and north-west, were worse affected than others.

'People were complaining pretty miserably and they couldn't understand why the heck the ESB couldn't provide electricity for the country,' O'Malley said.

In the medium term, the only realistic alternatives were coal or nuclear. O'Malley had already persuaded the cabinet to sanction an ESB coal-fired station at Moneypoint on the Shannon estuary in Co. Clare, at an outline capital cost of £350m, similar to the 1977 projected costs of a nuclear power station. Industry and job creation were ostensibly the more important part of his ministerial portfolio. By his own account, most of his time, particularly throughout 1979, was devoted to attempting to secure oil supplies for the Republic, including purchasing it directly, as well as chartering ships in the Gulf to bring the oil into the country.

O'Malley first attempted to acquire the rights to drill oil in a North Sea 'gold block', as the drilling blocks auctioned off to other countries by Norway in the late 1970s were known. On 20 June 1979, the minister announced that the government had agreed arrangements with the Irish-owned Aran Energy to take up a 51 per cent share in the company in the event of Norway agreeing to lease a gold block to Ireland. At the time, there were media rumours of Irish fishing rights being bartered to the Norwegians as part of the deal.[9] A basic requirement for the Norwegians was a national oil company. They refused to give a gold block to a private company and negotiations failed.

O'Malley first visited Baghdad in August 1979, some three weeks after Saddam Hussein had taken power, and although he did not meet with Saddam, discussions were held with the Iraqi Oil Minister and Tariq Aziz, who later went on to become Iraq's long-serving Foreign Minister. The Iraqis were prepared to sell 700,000 barrels of oil to Ireland at about $22 a barrel, subject to a number of conditions.

'The deal was done on a national basis, not on a commercial basis,' according to O'Malley, 'and I had to give them an undertaking that I wouldn't sell on any of the oil and that in particular I would ensure that none of it got into the hands of the Seven Sisters.'

On his return, O'Malley told the cabinet that Ireland would have to establish a national oil company to receive and process the Iraqi consignment. The Irish National Petroleum Corporation was formed.[10] O'Malley recalls that shortly afterwards he was approached by a major oil company, one of the Seven Sisters, and asked to divert the Iraqi oil. He was offered $7 dollars a barrel on top of what he'd already paid; a substantial profit margin. 'I told them I couldn't do it,' he said.

Within the cabinet, there was no serious opposition to pushing forward with the Carnsore project. 'As I recall I had no problem getting them to agree with whatever I put up,' according to O'Malley. 'Some had reservations because of the emotional carryover from Hiroshima and Nagasaki, that they were just unhappy or a bit uncomfortable, shall we say, with nuclear. But at the time, given the situation we were faced with, there was no alternative.'

O'Malley liked to tease his parliamentary opponents that he might reserve the final decision on a nuclear power station to himself as minister. He delighted in pointing out that the National Coalition government decision of 28 November 1973 had been made without any consultation with anyone, including 'Friends of the Earth or the recently distinguished winner of the Lenin peace prize, who feels very agitated that we should have a nuclear power station'.[11]

That decision still stood. 'Unless I want to go back and change what was done in November 1973 I need not go back to the Government at all,' O'Malley warned.[12]

He also felt himself entitled to some irritation at the cynical political posturing of the former National Coalition partners in face of the grim reality of Ireland's energy crisis. In opposition, Fine Gael had shifted their stance from nuclear necessity to not being opposed 'in principle' to nuclear power. Fine Gael identified what it regarded as the Achilles' heel of the nuclear project: the capital cost of the new station. Labour straddled both sides of the fence, marshalling all the arguments against development of nuclear energy in Ireland, ranging from a putative threat to Irish neutrality to the disposal of nuclear waste, while at the same time conceding in a half-hearted way the possibility that the nuclear option might, in the end, prove a necessary evil.

Introducing legislation to provide for ESB capital borrowing for the Moneypoint station, O'Malley expressed disappointment that only three opposition speakers showed up to address such a massive investment in coal-fired generation. And when they did, 'a great deal of the discussion of the three Deputies was related to the nuclear proposal'.[13]

O'Malley's policy on the nuclear option was in fact more flexible than his detractors gave him credit for. Far from how he was represented at the time by the anti-nuclear movement and subsequently by political opponents who never missed the chance to take a cheap political shot, O'Malley was no nuclear zealot. His, and the government's, short-term priority was to secure energy supply for a small country with distinct geographical and political disadvantages that inhibited the possibilities of interconnection with Britain, had no indigenous fossil fuel resources of substance and suffered from the absence of the economies of scale that applied in the energy market of larger countries. Diversification from dependence on oil was the key policy determinant, and coal and nuclear were the only viable alternatives.

As Minister, O'Malley sought to integrate long-term energy policy within his overall strategy for the industrial development of the Irish economy, the first time such an effort had been made. In June 1978, his department published a separate Green Paper on energy, which in simple yet erudite terms analysed all the options and invited the public to respond – an early and unprecedented, in Irish terms, exercise in public consultation.[14] Later that year, the Green Paper proposals were incorporated in the Government White Paper on Industrial Policy.

O'Malley had also instructed the ESB to engage with groups opposed to the Carnsore project and to provide comprehensive information to all responsible parties who requested it. In principle, O'Malley said, he had no objection to a public inquiry on the nuclear project, but he queried whether it would illuminate public debate or polarise it.

Ireland had no tradition of public inquiries and Britain's experience was not relevant to the Irish project.

'The inquiry which was set up last May into the proposal to expand for foreign purposes the Windscale reprocessing plant is the only one that was held that I or my Department can find in relation to any kind of nuclear plant in Britain. There were none held that I, at any rate, can discover in relation to any of the smaller stations,' O'Malley told the Dáil.

That inquiry, he added, 'cost hundreds of thousands ... if not millions of pounds' and it finally ended up with the judge 'telling the British Government that it was perfectly safe to go ahead with Windscale'.

'Within five minutes of the findings being given the people who had asked for the inquiry in the first place said that the results of it should be disregarded, and that it was all rubbish,' O'Malley noted. By inference, any Irish inquiry would meet a similar fate.[15]

Given the plethora of ESB studies and reports, the government considered that the main public concerns had already been answered. A report for the ESB by An Foras Taluntais, the Irish Institute for Agricultural Research, in December 1977, for example, comprehensively addressed the issue of the suitability of the Carnsore site, accident scenarios and the environmental impact in normal or accident conditions of the plant. The report, which included consultations with a wide range of experts including BNFL personnel at Windscale, assessed the likelihood of a major accident at Carnsore at about one in 400,000 years. It concluded there were no likely adverse effects to agriculture or human and animal health from the operation of the power station.

Anti-nuclear propaganda drew on a wider base, in which the Windscale accident featured prominently, conveying the impression that the Carnsore power station, in terms of relative safety, might turn out to be a replica of the Windscale piles. O'Malley's flat-earthers derived from many shades of political opinion within the country, including dissident republicans. Supported by international NGOs, the various components of the Irish anti-nuclear movement were already masters in the art of undermining public trust in official information.

* * *

Initial reaction to the Harrisburg accident in Ireland was muted by comparison with the public demonstrations and demands for abandonment of nuclear power that had caused political uproar in other European countries already with nuclear programmes. As a 'first-time buyer' in the nuclear marketplace, the ESB adopted a wait and see approach, awaiting the outcome of the official investigations into the accident and then seeing what modifications might be entered to its own specifications to achieve reactor safety design goals. The government dispatched Frank Turvey, a senior scientific officer with the Nuclear Energy Board, to attend the official investigation into the TMI accident at Harrisburg.

Turvey, one of three senior officers who had joined the staff of the NEB in 1976, coincidentally shared a similar technical background – expertise in nuclear submarines – with several of the operating staff at the ill-fated TMI-2 station. He had joined the Royal Navy as a cadet in the early 1950s, leaving an engineering course in UCD to follow in the footsteps of a family tradition that included his own father and three of his uncles on either side of the family. One of them, George Firth, from Waterford, was known as 'the President's captain' on account of his adventures ferrying Eamon De Valera, both officially and incognito, across the Irish Sea in the early 1920s, including following De Valera's celebrated escape from Pentonville prison.[16]

Prior to joining the Royal Navy where he would spend twenty-three years, most of it on Polaris submarines, Frank Turvey had consulted Fr. Rupert Coyle, one of his old teachers in Belvedere College in Dublin. Fr. Coyle advised the young would-be cadet that Soviet communism was the greatest threat facing the Western world and, in such circumstances, joining the British navy was not a sin.

By 1977, the NEB had produced its first order to regulate and control the use of radioactive devices by hospitals and industry and within a year had 150 licencees on its books. Turvey's NEB colleagues, Noel Nowlan and former ESB engineer John Cunningham, presided respectively over the project to establish a laboratory and monitoring system to measure radioactive contamination of the Irish Sea from Windscale and the licensing and use of radioactive sources nationally.

The NEB had a total staff of nine but its resources were stretched and its funding, a mere £153,000 in 1977, was pitifully small to the magnitude of its task. In the view of its own staff, the NEB could never meet the expectations of the legislation that had established it, because the Department of Finance never allowed it sufficient resources to do so.

As the government awaited the verdict of its NEB and ESB experts on the Temeny inquiry, set up to examine the TMI accident and whose report was published in late 1978, as well as other investigations into the accident, department officials set about devising a framework for decision making on the Carnsore project.

A government decision, announced on 24 February 1979, instructed the ESB to proceed to final specification for the Carnsore power station. It established an interdepartmental committee to consider all aspects of the project. New government legislation on planning was promised. In effect, the government proposed taking the planning decision away from the local authority, Wexford County Council, and from the national planning board, An Bord Pleanála. Instead, the planning case for nuclear power in Ireland would be considered as part of a broad-ranging public inquiry whose report would be published prior to a final government decision on the Carnsore project. The current estimate for the capital costs of the plant had risen to £400m,[17] but in the midst of a severe energy crisis, this appeared a reasonable investment to secure an independent energy supply in the medium term.

Conceding a public inquiry failed to take the heat out of the issue, as the media and the anti-nuclear lobby persisted in questioning the likely independence of any inquiry. A growing number of deputies in all political parties began to appreciate the self-promotional value of newspaper headlines taking a critical stance on the nuclear issue.

O'Malley and his pugnacious junior minister, Raphael P. Burke, neither of them disposed to suffer fools gladly, if at all, made a formidable parliamentary team. Ray Burke had little or no direct involvement in the energy portfolio and was instead designated by his senior minister to look after trade missions and the commercial sections of the departmental brief. Nonetheless, when the ambitious young Fine Gael deputy for Meath, John Bruton, repeatedly raised the safety of Wylfa power station in Anglesea, 'only 60 miles of uninterrupted sea from the city of Dublin', and demanded a permanent monitoring presence by the NEB 'to ensure that there is safety', it was Burke who was sent in to deal with him.

Safety assurances from the British authorities could not be accepted 'on faith', Bruton alleged, since the British 'obviously have an interest in minimising any concern which might be brought to notice because they have already invested a substantial amount of money in these reactors and do not wish to be proved wrong'.

He was insistent that even low-level discharges to the Irish Sea could pose a risk to the health of Irish people.

'While no individual discharge may be dangerous, the accumulation of discharges over a period could create a dangerous situation,' he opined. 'The discharge of atomic waste from the Windscale nuclear plant into the Irish Sea is increasing the risk of cancer among people who consume fish.'

Burke was having none of it. 'It is nice to know that the Deputy has discovered that there is a nuclear plant at Wylfa eight years after it came into operation,' he replied. 'He says he is not qualified personally to make any comment on it. Having listened to his contribution I can confirm that he is not qualified personally to make any comment whatsoever.'

Burke branded Bruton as irresponsible. 'It is disappointing, to say the least, that the Deputy should by innuendo try to terrify people by taking advantage of his position in this House,' he admonished.[18]

O'Malley could also count on support for his energy policy from the Taoiseach, Jack Lynch, to whom he was both personally and politically close. Returning from a Brussels summit meeting in June 1979, the Taoiseach was reported by the *Irish Press* as saying that when it came to nuclear power: 'We were advised very strongly to proceed at the quickest possible pace.'[19] Fine Gael leader, Dr Garret FitzGerald, accused the Taoiseach of disregarding the safety of the Irish people and 'rushing into the idea of a second nuclear reactor before there is an inquiry into having any nuclear reactor'.[20]

* * *

Lynch's tenure as Taoiseach was doomed. Two years into the lifetime of his government, the mild-mannered Corkman, the most popular political leader ever in the state's history, was under pressure. The hollowness of his government's promises of prosperity was exposed by static economic growth, rising unemployment, escalating inflation and growing concern about the extent of government borrowing.

The Fianna Fáil party, for the first time ever, was openly split over economic management and Northern Ireland policy. Lynch had been the compromise leadership successor to Seán Lemass. The political trophy of Lynch's stunning electoral victory of 1977 was turning into more a poisoned chalice as many of the new young Fianna Fáil deputies began to fear they might lose their jobs come the next election.

Looking around for a prospective leader outside the Fianna Fáil old guard, most would ultimately pledge their allegiance to Charles J. Haughey, whose leadership ambitions had been thwarted by Lynch's accession in 1965, but whose thirst for power, despite several notable upsets in his political career, had remained undiminished over time. As Haughey remarked to a journalist at the Dublin count centre on the night the results of the 1977 election poured in: 'Those are all my men.'[21]

Charles Haughey became leader of Fianna Fáil in December 1979. His rival for the leadership, George Colley, was made minister of a new Department of Energy, hived off from Industry and Commerce. O'Malley advised the new Taoiseach to separate the two departments. At first, the Taoiseach offered the new department to O'Malley. But he refused it.

'So he then gave it to Colley. I think Haughey thought it wouldn't be a very important department. He didn't want Colley making headlines. There was huge animosity between them,' O'Malley said.

According to Dr Martin Mansergh, advisor, friend and political confidante of Haughey over several decades: 'One of the things, indeed maybe the only one thing, that Colley and Haughey agreed on in 1980 was to kill off the nuclear project and that was announced pretty quickly.'

No such agreement was ever adverted to publicly by the new Minister for Energy throughout his period in office, though the Carnsore project steadily slipped down the government's order of priorities.[22] George Colley delayed introducing legislation to set up the promised public inquiry. Until the interdepartmental committee published its report, the minister could see no reasons for 'rushing ahead with legislation'.[23]

Mansergh asserts Haughey took 'quite a pronounced anti-nuclear stance from the outset' and opposed 'both the military and civil uses of nuclear power' Mansergh, in his career as an official in the Department of Foreign Affairs, had represented Ireland on several international nuclear bodies and for a time was posted in Germany, when the debate on nuclear power in that country was at its height. As a godson of Canon John Collins, co-founder with Bertrand Russell in 1958 of the Campaign for Nuclear Disarmament, Mansergh claims Dr Collins exerted a major influence on his thinking about nuclear issues, particularly throughout the 1980s. Mansergh was appointed to the interdepartmental committee set up in February 1979. As the meetings progressed, his attitude to the nuclear project changed from 'supportive to sceptical'.

O'Malley's recollection is that Haughey had taken little or no interest in the nuclear power proposal at cabinet when it was seriously discussed in 1978 and 1979. There was nothing in Charles Haughey's parliamentary or public record to indicate any principled objection to nuclear power until later in the 1980s when it became a populist issue.[24]

By the end of 1980, the urgency had gone out of Ireland's energy crisis as oil supplies became easier to find. O'Malley recalls that Colley advised the cabinet towards the end of that year that there was no need to push ahead with the Carnsore project. 'And the government agreed with that,' O'Malley said.

George Colley announced that a final government decision on the project was postponed for a further two years.

In a latterly infamous televised address to the Irish people on 10 January 1980, Charles Haughey confirmed that the country had long been living beyond its means. Frank Turvey's ultimate assessment of the implications of Three Mile Island for Ireland's nuclear ambitions – that containment at the TMI-2 reactor had worked to save the environment but economically, any such accident to an Irish based reactor would be a disaster – was by now academic. No Irish government facing into the 1980s could afford to risk creating a potentially massive technological white elephant.

The repeated destruction by the IRA of the North–South interconnector[25] from 1975 onwards removed the rationale for an all-Ireland nuclear energy project, as well as the prospects for maintaining grid stability while operating a 500 MW nuclear power station economically. The Programme for Government of the short-lived Fine Gael–Labour coalition of 1981 promised that: 'No nuclear power station will be approved as long as any doubts remain about the environmental hazards or waste disposal links of such a project.' The favoured political option of all parties was to leave the Carnsore project to wither on the vine.

PART FIVE

IRELAND V BRITAIN: THE SELLAFIELD PHONEY WAR

15

Atomic Village

'I do not know if many Deputies experience as I do a very real feeling of anger when I consider the enormity of this situation and what we in this country are being asked to tolerate.'

Charles J. Haughey, March 1986

On a clear day, back in the 1970s, the physical outline of the Sellafield site, the cooling towers of Calder Hall and the tall chimneys of the Windscale piles were all plainly visible from the beach at Seascale, three kilometres away, and from most other vantage points in the village. Seascale has its own stone circle, the remnant of a Bronze Age settlement, demolished in the mid-nineteenth century by a local farmer who buried the stones. Later, the stones were resurrected and remounted in their original formation as a tourist attraction. Like other coastal towns in nineteenth-century Britain, Seascale had flourished as a seaside resort, especially after 1850 when it was first connected by rail to Whitehaven to the north and Barrow to the south.

The village comprised a few shops, a hotel, some boarding houses, a Methodist chapel, a church, a small number of private houses and a Church of England school. Prior to the Second World War, Seascale had a population of about 690 people and was popular with middle-income families taking summer holidays. 'Nanny took the children on the sands and father played golf while mother read a novel or, alternatively, the parents took the day off and explored the mountains and lakes. The commercial and social life of Seascale centred around catering for such people,' Hugh Gethin Davey wrote of it.[1]

The traditional coal and iron ore industries of West Cumberland took the full blast of the 1930s depression in Britain. Like much else in the county, Seascale's tourist trade too fell into decline. After the war, there seemed little hope that Seascale's fortunes as a middle-class holiday haven might ever recover.

While the Windcale project was at the planning stage, the Ministry of Supply considered the option of building an 'atomic village', separate to existing local communities. An estimated 1,000 skilled and professional workers needed to be brought into a community that itself lacked the requisite skills to staff the nuclear complex. It was essential to house many of these professional staff in reasonable proximity to the site, in case of plant emergencies. Seascale had been relatively unaffected by the temporary influx of workers to man the ordnance factories at Sellafield and Drigg during the war, most of whom had taken up residence in larger towns north of the site, like Egremont, ten miles away, or in Whitehaven. In 1947, the Ministry of Supply placed a contract with the Whitehaven local authority for the erection of 300 new houses at Seascale.

Seascale's two-roomed village national school was designed to accommodate about forty pupils; but by 1950 that number had risen to 105. A year later, Seascale had a new school and the population of the village mushroomed to about 2,000 people. For the next several years and beyond, the population of Seascale remained in considerable flux, sometimes rising to over 3,000 depending on the level of construction activity at Sellafield, and as middle-class families moved in, or just as frequently, out to other towns in the area.

Seascale residents had mixed feelings about the massive nuclear site on their doorstep. As its distinctive skyline took shape and form, there was an inevitable realisation that their village could never be the same again. Some had genuine misgivings about the construction of the 300 new houses at Seascale. The new-comers, the majority of whom were from the professional classes, however, threw themselves into village life, reviving the local horticultural society, engaging in Church activities and establishing a works social club that welcomed the original inhabitants as associate members. In time, Seascale became just another West Cumbrian town, most of whose residents depended on the atomic plant for their livelihood, and whose immediate living environment was enhanced by the beautiful, golden sandy beach that extended for miles along the shore line.

* * *

Only fifteen curies of Iodine 131 were released during the Three Mile Island accident, one-thousandth of the amount released during the Windscale fire and a tiny fraction of the amount in the Chernobyl accident seven years later.[2] TMI, however, had fuelled a furious argument among western scientists about safe levels of radiation, and whether or not there was a 'threshold dose' below which individuals exposed to radiation experienced no harm. Until the mid-1950s, when public controversy grew about the health impact of fallout from atmospheric weapons' tests, the threshold dose was largely accepted among scientists, despite some pioneering studies that suggested the possibility of long-term damage even at very low doses of radiation.

Radiation affects the body by displacing electrons in individual molecules and thereby altering their chemical composition. In very high doses of radiation, too many cells are either completely destroyed or damaged to effectively repair themselves. Anyone exposed to such high doses can be expected to die of radiation sickness within a period of hours, days or weeks depending on the extent of irradiation of the body, in what is known as the 'deterministic' effects of radiation. Exposure to much lower levels of radiation also inevitably damages cells, but the damage is naturally repaired and no individual adverse effects are experienced in the short term. Cell repair may, however, be incomplete, leading to a greater risk of cancer in later life or of the passing on of inherent cellular defects to succeeding generations, the so-called 'stochastic effects' of radiation. For the purposes of determining dose limits both to the public and to radiation workers generally, the International Commission for Radiological Protection (ICRP) assumes there is no dose of radiation below which no risk occurs. This is known as the linear non-threshold theory (LNT).

From the point of view of science, this does not, however, equate with the ridiculous proposition that all radiation doses are therefore harmful, since natural background radiation – from rocks (radon gas) or the sun (cosmic radiation) – affects all living organisms and obviously radiation is essential to life, sunlight being the most obvious example. The purpose of LNT is to devise a regulatory framework within which safe dose limits of high-energy ionising radiation for nuclear workers and the public can be established. Like all areas of scientific investigation, dose limits, and the notion of the LNT itself, are constantly subject to change in light of further studies on the effects of radiation.

Studies of the radiation dose to the 200,000 or so inhabitants within a twenty-mile radius of the TMI plant consistently showed that individual doses were about one-sixth that of a normal chest X-ray and well below the levels of normal background radiation in the area.

The studies also revealed a new phenomenon, 'nuclear neurosis', which led people in the area to attribute the cause of all illnesses, as experienced by themselves, members of their families or farm animals, to contamination received at the time of the accident. Expert studies that discounted such empirical accounts were dismissed by many as part of an official cover-up.

The Harrisburg accident revived interest in the release from the Windscale fire of 1957, as well as coinciding with the imminence of the twenty-fifth anniversary of the accident. Britain's National Radiological Protection Board (NRPB) decided to reassess the health impact of the radiation released during the fire.

Peter Taylor of the Political Ecology Research Group in Oxford was commissioned by the US Union of Concerned Scientists in Massachusetts to conduct a new study. Taylor's report in July 1981 suggested the Windscale fire would cause as few as twenty or as many as 280 deaths from various cancers in the United Kingdom and Western Europe over several decades as well as a probable 250 cases of thyroid cancer directly related to the iodine 131 released during the fire, of which ten to twenty-five cases might prove fatal. Taylor's report also mentioned leukaemia clusters, which he felt were worthy of study though unrelated to the Windscale fire and its aftermath. The following year, the NRPB published the Crick and Linsley study, which was confined to the United Kingdom and estimated the likely number of fatal cancers at twenty.[3]

The Crick and Linsley study, however, had failed to take account of releases of polonium 210 or emissions of small quantities of tritium, plutonium and uranium during the fire. This omission was seized on by the press. Polonium 210, it transpired, accounted for 36 per cent of the total dose to the public from the fire's emissions, second only to iodine 131 at 39 per cent. Crick and Linsley published an addendum to their report the following year, increasing the upper bound of fatal cancers arising from the Windscale fire to thirty-three.

The scale of the Windscale release had remained more or less under scrutiny over the years since the fire. But the studies were the stuff of scientific journals and rarely attracted wider public attention. Any mention of materials such as polonium 210 in early calculations of the release inventory were cursory and probably included only because traces of polonium 210 had been detected in the fallout over Europe. Unwillingness to draw attention to the presence of polonium in the release was later explained away on the basis of politics: UK governments in the 1950s were reluctant to disclose to the US that polonium 210 was being manufactured in Windscale on any significant scale since its only possible use – as a chain reaction trigger in atomic weapons – would only show how far behind the Americans the British weapons' programme was.

It was a Yorkshire Television (YTV) programme, 'Windscale – the Nuclear Laundry', broadcast in November, 1983, that finally took such considerations of high politics out of the representation of hard facts and brought the industry and government down to earth. The programme presented details of a cluster of leukaemia and non-Hodgkins lymphoma among children in Seascale between the years 1954 and 1983. In the absence of any other obvious cause, the finger of blame pointed squarely at the site.

* * *

The YTV team approached the Sellafield management requesting co-operation on a planned documentary programme on the health of radiation workers in 1983.[4] The issue had achieved a brief notoriety a few years earlier, in 1975, when the deaths of two former Windscale employees from cancer were reported in the national media.[5] The Sellafield management told the film crew there was no definite indication of any long-term ill effects among its workforce from radiation exposure. This was not a matter of opinion but established fact as far as the site management was concerned: BNFL had financed its own study on workers' health, published in 1983,[6] whose findings were later broadly endorsed by an independent study of the workforce in 1986.[7]

Rather than fight claims alleging radiation-related cancer deaths on a case by case basis, BNFL in 1983 had introduced a general compensation scheme, which provided awards on a sliding scale in proportion to the likelihood of occupational radiation exposure having contributed to the death. In return, recipients would not pursue BNFL in the courts.[8]

In the course of his visit to the area, the programme producer James Cutler picked up rumours of cases of childhood leukaemia in Seascale. On 1 November 1983, Yorkshire TV broadcast its prize-winning documentary on nine cases of childhood leukaemia and non-Hodgkins lymphoma that had occurred in Seascale in the period since 1955, an incidence apparently ten times the national average for a community of that size.

BNFL's initial reaction was defensive. As Director of Health and Safety, Peter Mummery was charged with the task of making the company's response on television.

'He looked like Adolf Eichmann being interrogated about the extermination camps,' a former colleague later ruefully recalled. 'All you see is Peter Mummery, fiddling with his pencil continuously while he's talking, and then being quite aggressive, and all you think is this man has clearly got something to hide.'

The parade of media consultants that thereafter regularly trooped in to BNFL to sell their wares highlighted Mummery's programme clip in their presentations, as the perfect example of how not to behave on television.

Part of Mummery's problem was that the existence of the cluster had taken the medical authorities, and BNFL, completely by surprise. An even greater part of it was that he came from the old school, the indigenous nuclear lobby embedded within the political establishment that was the UKAEA, which for over twenty years held a monopoly on the nuclear policy advice provided to successive British governments; advice that passed without critical scrutiny either by parliamentary institutions or a quiescent media.

A British debate on nuclear energy had only begun to gather momentum at national level following BNFL's application for planning permission for the Thorp plant in 1975, now generally accepted as a tipping point in public perceptions of civil nuclear power. The plans for Thorp gave Friends of the Earth a national platform. Securing the first public inquiry into a nuclear project in Britain, the Windscale Inquiry of 1977 was a major strategic victory for the fledgling UK anti-nuclear movement. However disappointed the campaigners may have been in Judge Parker's report, the shutters of secrecy that had previously shielded the nuclear industry from public scrutiny were torn away. The Thorp project remained pivotal to continuing anti-nuclear campaigns by Friends of the Earth and, later, Greenpeace UK, established in 1977.

Newspapers and television cast themselves as the forum for a national debate on nuclear energy, and for calling the industry to account. The media, ill-informed as individual journalists may have been about the issues in their own right, were set to lead public opinion, always susceptible to scary headlines about any new hazards or risks emanating from an industry of which they understood little.[9] An increasingly negative media portrayal, reinforced with periodic bursts of outrage over reports of accidents and incidents at Sellafield itself, had brought about a fundamental shift in public support for nuclear power by the early 1980s. General public support for new nuclear power stations in Britain had registered at 56 per cent in opinion polls in 1978. In 1981, polls showed public opinion was 55 per cent against.[10]

In Thatcher's Britain, the Prime Minister was herself labelled a champion of the nuclear industry, especially by the Labour Party and trade unions who believed her support for a new programme of nuclear power stations was intended primarily to curb the traditional industrial muscle of coal miners in the energy sector. In 1982, Margaret Thatcher's first government proposed building ten new PWRs, an expansion of 15,000 MW on existing nuclear capacity. Significantly, the government also honoured the promise of the previous Secretary of State for the Environment, Labour's Peter Shore, by granting a public inquiry into the first of the proposed stations, Sizewell B, at Suffolk.[11] The inquiry, from January 1983 to March 1985, would run for over two years, once again ensuring sustained media attention.

Thatcher's government responded to the Nuclear Laundry controversy by immediately appointing a committee of inquiry into the Seascale cluster, chaired by Sir Douglas Black, the renowned medical researcher whose earlier work had included a seminal report on the relationship between poverty and ill-health in Britain.[12]

A fundamental problem with the Yorkshire TV programme was that it used the existence of the cluster to fit a pre-determined conclusion: that Sellafield radiation was to blame. It was, one commentator later suggested, like firing a bullet at a blank wall and then drawing the bullseye around the hole it had made. Black's report in 1984 found no evidence of abnormal levels of cancer in West Cumbria, but twenty-two cases of leukaemia and lymphoma among under-twenty-fives living to the south of Sellafield had occurred since 1955, and the excess was concentrated in Seascale. The abnormally high number of cases of childhood leukaemia in Seascale merited further examination, Black concluded.

The National Radiological Protection Board (NRPB) had provided evidence to Black showing radiation from Sellafield was only a fraction of that received from normal background radiation. There was a suggestion that the pre-1957 discharges from the Windscale piles might bear some responsibility, but most of the Seascale cases had neither been born nor conceived when that contamination occurred.

It was a mystery as to why Sellafield radiation should be selective, only affecting children. As other clusters were reported around nuclear sites throughout the country in the years that followed, including Dounreay in Scotland, the popular impression that radiation was responsible was sustained, while for scientists the puzzle only deepened. The discovery of a cluster in the small community of Largo Bay in Scotland was even more perplexing – Largo Bay had been selected as a possible site for a nuclear power station, but none was ever built. Scientists, including the internationally renowned Richard

Doll,[13] began to postulate that some other factor – possibly a mystery virus – was at work.

In 1988, the epidemiologist Leo Kinlen reported that a cluster of childhood leukaemia had occurred in the Scottish new town of Glenrothes. Kinlen suggested that population mixing was probably a causal factor, triggering an unusual response to a common infection among the affected population. Later, he went on to identify childhood leukaemia clusters near army bases, large rural construction sites and in the areas to which wartime evacuees had been sent. If Kinlen's theory was right, the expansions in Seascale's post-war population and the constant movement and mixture of people from different parts of the country bearing different pathogens, could account for the excess in childhood cancer cases in the village over a thirty-year period. It would take another two decades before Kinlen's theory, as it became known, was finally acknowledged as substantially correct.[14]

Black's report recommended more effective registration of cancer cases in Britain and tighter regulation of emissions and discharges from nuclear installations, recommendations duly accepted by the government. Black also advised new lines of inquiry into cases of leukaemia and lymphoma diagnosed in people under twenty-five years of age living in West Cumbria, and investigation of the records of all children since 1950 who had attended school in the village or were born to mothers who had lived at Seascale. The government-sponsored Committee on Medical Aspects of Radiation in the Environment (COMARE) was established in November 1985 in response to the Black Report's final recommendation.

In the quest to discover the cause of the Seascale leukaemia cluster, over the next twenty-five years COMARE, along with various teams of independent researchers, would generate a vast library of epidemiological and medical data on Britain's nuclear workforce and local communities. Some, like the Gardner report in 1990 with its initial findings of pre-paternal irradiation as a factor in subsequent leukaemia among the children of radiation workers, proved hugely controversial and, for a time at least, appeared to pose a direct threat to the survival of West Cumbria's main industry.

* * *

Local confidence in Sellafield received a setback in October 1981 when it was disclosed a release of iodine 131 had contaminated milk supplies on two farms in the immediate vicinity of the site. The release,

300 times greater than normal, was caused by reprocessing a consignment of spent fuel that had not been left to radioactively decay for a sufficient period of time at the original power station before being shipped to Sellafield. Details of the release were not published by the Sellafield management for several days, by which time milk from the farms had already been sold and consumed. Then, just fourteen days after the Nuclear Laundry broadcast, a far more serious incident occurred. Coincidentally, four Greenpeace divers were on hand to bear witness to it.

The irony of the Sellafield beach incident, as it came to be called, was that it happened at a time when the site management was progressively reducing the radioactivity of discharges to the Irish Sea from Sellafield. The site ion exchange plant, SIXEP, an effluent treatment plant that removed caesium from pond water and other liquid wastes that arose from magnox reprocessing, had come on line in 1975. The beach incident would further dramatically increase the use of the best available technology to remove radioactivity from discharge wastes and reduce emissions from Sellafield plants. In investment terms, BNFL spent £2bn on clean-up technologies over a twenty-year period: the end result was a reduction in radioactive discharges by the mid-1990s to less than 1 per cent of peak levels in 1974–5.

The beach incident resulted from management error. Magnox reprocessing was shut down for annual maintenance and the plant was flushed out in or around 11 November 1983. The waste liquid from this operation, containing possibly as much as 4,500 curies of radioactivity suspended in solvents and particulate matter, known as crud, was directed to one of the sea tanks normally used to hold low-level liquid wastes for final checking before discharge to sea. Once the mistake was discovered, the operators first attempted to pipe the contents of the sea tank to another holding tank. When they realised this transfer could take several days, and considerably delay the maintenance schedule, they reconsidered their options. Since discharge of the material in the sea tank should not exceed the normal site authorisations, they decided to let it go to sea in the hope and expectation that rough waves might disperse the material quickly and very little of it, if any, would find its way back to land. As it happened, the sea was unusually calm.

BNFL's press office at Risley had already been alerted by the *Guardian* that the paper was researching a Greenpeace story that alleged four of its divers, who were taking samples from the end of the Sellafield pipeline, and the rubber dinghy they were using, had been contaminated on the night of Monday, 14 November. Four days later, a senior manager at Sellafield phoned Harold Bolter to say he could see an oily slick off the coast from his office window on the site.

'I got the distinct impression that there had been similar discharges of solvent and crud in the past,' Bolter later wrote, 'but the strong tides and heavy swell of the Irish Sea could normally be relied upon to carry the material away from the shore and disperse it.'[15]

Whatever his impressions and his anger that the Sellafield management had not informed corporate HQ about the incident for over a week after it had begun, Bolter had no choice but to report the event to ministers under the criteria previously laid down by Tony Benn. Within days, solvent and particles washed up on the beach and the Sellafield constabulary warned local residents of Seascale to stay away and not allow their children to pick up anything from the shore.

Bolter issued statements asserting there was no risk to public health and safety. The first statement on 18 November simply admitted to a discharge containing 500 curies of radioactivity. The next day the company issued an 'all is well' statement, claiming that full public access was now restored to the local beaches. However, the Department of Environment warned people against 'unnecessary use' of the beaches, a ban later extended to thirty miles along the coast, lasting for over several months until well into the summer of the following year.

In the House of Commons, a furious government announced the matter was being referred by the regulatory authorities to the Director of Public Prosecutions. The local Labour MP, Jack Cunningham was, if anything, even more infuriated by BNFL's delay in announcing the incident, their flawed and inaccurate communications and the understandable ire of his constituents, especially those in his local community engaged in the tourist trade. The Sellafield workforce, who felt their families and children were directly affected by the beach ban and most at risk from any radiological consequences of the event, rounded on the local management and began to organise their own lobbying movement to protect their jobs and their industry.

BNFL was eventually fined £10,000 plus costs when the case finally came to court in June 1985, the first time a British nuclear company had faced criminal charges. The real cost was much higher, particularly in enforced expenditure on new waste treatment facilities, including construction of an enhanced actinide removal plant, EARP, designed to take plutonium, americium, and other heavy radionuclides out of all future sea discharges. The worst damage, however, was the almost irretrievable loss of BNFL's corporate reputation in all the places where that mattered.

* * *

BNFL had engaged in a running battle with Greenpeace from the late 1970s, first over the use of the port in Barrow-in-Furness for imports of spent fuel from Japan and Europe and later attempts by Greenpeace divers to block the Sellafield discharge pipe. High-profile stunts by Greenpeace, including blocking the harbour entrance to Barrow, abseiling onto and chaining themselves to ship decks, or dangerously running inflatables in the path of oncoming vessels in the Irish Sea provoked massive media coverage but made little or no impression on UK government policy. BNFL retaliated by taking out court injunctions that resulted in large fines to Greenpeace. Aware this strategy was a two-edged sword, since public sympathy gravitated more towards an environmentally committed, international voluntary organisation than an obscure, notoriously secretive, nuclear services company in England's North West, BNFL usually held back from pursuing Greenpeace for costs.

Greenpeace's declared intention in 1979 to block the Sellafield discharge pipe provoked an application for a permanent injunction. Greenpeace's first attempt, in 1984, left them with a penalty of £50,000 plus costs, which the organisation could ill afford to pay.[16]

Ireland presented a back door through which Greenpeace could mount political pressure on the British government. Like their British counterparts, the mainstream Irish media was by now almost universally anti-nuclear in its editorial stance, with much coverage devoted to the dangers to the health and well-being of Irish people living on the east coast from Sellafield discharges to the Irish Sea.

Raising the nuclear bogey proved irresistible to politicians right across the political spectrum in early 1980s Ireland beset with economic stagnation, rampant inflation, an unemployment crisis and an atmosphere of political instability that saw three changes of government in as many years. An element of mistrust had entered political discourse on nuclear issues, partly a reflection of the popular mood which had turned against any nuclear project in Ireland, but also relating to the sharp downturn in Anglo-Irish relations that followed the beginnings of the Thatcher era in Britain.

With tenacity and persistence, the Irish governments of the early 1980s, led by Charles J. Haughey and Garret FitzGerald respectively, had sought to convey an understanding of Northern Irish politics from a nationalist perspective to an unsympathetic Margaret Thatcher, before and after the disastrous IRA hunger strike campaign in Northern Ireland's Maze prison that resulted in eleven deaths of the hunger strikers themselves during 1981 and wider carnage amongst the innocent civilian community.[17] Relations with Britain were further strained by Haughey's refusal as Taoiseach, alone among European Community member states, to support the established EC line on the hundred-day Falklands war in 1982.

The Irish government was reminded that 'Britain no longer rules the waves',[18] as deputies urged the FitzGerald-led Fine Gael–Labour coalition that took office later that year to make representations directly to British ministers about dumping nuclear waste in the Atlantic, the increasing number of Sellafield mishaps and continuing discharges to the Irish Sea. Fianna Fáil's Ray Burke, now in opposition, queried the source of Nuclear Energy Board advice that Sellafield discharges posed no risk to the Irish population.

'The information the Nuclear Energy Board are relying on is supplied by British Nuclear Fuels,' he said. 'They were found out telling lies regarding emissions from Windscale in the past. Nobody has any confidence in British Nuclear Fuels.'[19]

At first, Nuclear Energy Board staff adopted a policy of correcting what they regarded as the more outrageous examples of misinformation in the media, either through seeking interviews on radio and TV or in letters to newspapers. But the NEB board soon decided that the agency was at risk of being labelled as 'pro-nuclear' both by the general public and politicians. The policy was abandoned, to no good effect since Dáil deputies and some journalists in any case regularly accused the NEB of being pro-nuclear.

Press reports of the Nuclear Laundry programme, followed shortly by the beach incident, had entirely dissipated whatever public or political trust remained in British nuclear assurances. The health issue acquired a distinct local dimension in November 1983 when Drs Patricia Sheehan and Irene Hillery reported a cluster of Down's syndrome children born to six mothers who, the researchers claimed, were pupils at the St Louis convent in Dundalk in October 1957. At the time of the Windscale fire the school had experienced an outbreak of the Asian 'flu sweeping across Europe and there had been unusual downpours of rain.

'We are left with a nagging doubt that possible exposure to radiation associated with some infection had an adverse influence on the subsequent non-disjunction of chromosome 21 in their six babies,' Dr Sheehan concluded.[20]

* * *

Government ministers used the NEB as a shield when faced with what they habitually dubbed 'hysterical' challenges from the opposition and some of their own backbenchers about Sellafield and British nuclear issues. A nuclear regulatory agency was a requirement under the Euratom Treaty for Ireland's accession to the European Community. At first, the NEB was envisaged as a licensing body and overseer for Ireland's own nuclear project. As Carnsore's prospects faded into the distance the NEB's public information and radiological protection functions came more into prominence.

It was all very well for politicians in Dáil Éireann to demand the NEB engage in independent monitoring of Sellafield discharges to the Irish Sea; in government the same politicians were unwilling to provide resources for the NEB to install a national radiation monitoring network or establish a properly equipped laboratory to analyse samples.[21] Laboratory facilities at Trinity College and University College Dublin were far superior to any available to the NEB.

For decades, the NEB and its successor, the Radiological Protection Institute of Ireland (RPII), would plead for the establishment of a secure repository for Ireland's own nuclear wastes, sealed sources used to calibrate factory equipment, radionuclides used in hospital treatments or simple devices like smoke alarms and lightning conductors, that could not be repatriated to the country of origin, mainly the UK.

Much hospital radioactive wastes simply ended up in the sewers. Hospitals and laboratories throughout the country established on-site temporary storage facilities for solid nuclear wastes. Unwanted nuclear sources were accidentally sent to landfill, others abandoned in factories long since closed, or sometimes left sitting for years, forgotten by those who had used them. One radioactive source went temporarily missing in a stolen van.

In time, the NEB insisted on 'return to sender' contract clauses for new sources coming into the country. Most nuclear waste of this type that was returned to the UK ultimately ended up at Drigg.

On Windscale discharges, the NEB developed solid working relationships with their counterparts in Britain, including participation in a sea-sample monitoring exercise with MAFF, the Ministry of Agriculture, Fisheries and Food, and the conclusion of an agreement with the Nuclear Installations Inspectorate in 1980 to provide rapid notification of any incidents at British nuclear installations or of any other major developments in the industry. Frank Turvey was appointed as an observer by the OECD's Nuclear Energy Agency of nuclear dumping missions in the Atlantic in the late 1970s, about 500 miles off the south coast of Ireland, and advised on Ireland's resolution at the London Convention to ban nuclear dumping at sea in 1982. His colleague, John Cunningham, was one of the main advisors to government on Irish submissions to the Oslo/Paris Conventions on marine discharges to the North Atlantic. The NEB sent an observer to the 1977 Windscale Inquiry.

In the bleak economic atmosphere of the 1980s, any more than for any other state body, there were no extra resources available for the NEB to properly fulfill its radiological protection mandate. The NEB was particularly concerned at the absence of any fully developed nuclear emergency plan for the country. The agency's weakness was finally and brutally exposed on a weekend in early May 1986, when the fallout from the Chernobyl accident rained down on Ireland.

16

Close Down Sellafield

'The cry of my baby sister born yesterday calls out to heaven for action so that the 'hard rain' will never fall again.'

Letter from eleven-year-old boy in Co. Wicklow
to the *Irish Times*, 20 May 1986

A five megawatt reactor was connected to the grid at Obninsk, about sixty miles south of Moscow, on 26 June 1954. The reactor, the first in the world linked to a national electricity network, generated enough power for about 2,000 homes. It was of exquisitely simple design with relatively inexpensive components – graphite blocks that acted as a moderator with pierced channels for very slightly enriched uranium fuel, ordinary water to act as a coolant and generate steam for the turbines – a precursor of the RBMK, the 1,000 MW reactor that disintegrated at Chernobyl in the Ukraine, at 1.23 a.m. on 26 April 1986 following an ill-fated engineering experiment.[1]

The nearest Western equivalent to the RBMK was the old US Hanford reactors in Washington State. As nuclear reactors go, the RBMK was cheap to build and highly efficient in the production of plutonium, the main reason why of the 17 RBMKs built were all confined to Russian territory. But the design was inherently flawed: it incorporated what engineers term a 'positive void coefficient'. In a given set of circumstances while operating at lower temperatures the reactor was prone to sudden power surges, a runaway chain reaction liable to overheat the fuel rods to the point of explosion.

Unit 4 at Chernobyl was designated for experimental tests on the night of 25 April 1986. In the twelve hours preceding the accident, most of the standard safety systems that might have protected the reactor from an uncontrollable power surge, including the control rods, were either disabled or circumvented by the operators in order for the tests to be completed on schedule. When the power surge, a hundred times greater than normal, occurred within four seconds at

1.23 a.m., it was already beyond the operators' control. Unit 4's thermal energy increased from 200 MW to about 360,000 MW. The first explosion is believed to have been caused by the interaction of steam with disintegrating hot fuel elements in the reactor core. Seconds later it was followed by a second, hydrogen-based, explosion.

The reactor core was destroyed and its 1,000 tonne concrete cover blown off, jettisoning chunks of burning graphite and pieces of fuel into the immediate vicinity and propelling millions of curies of radiation in gases and aerosols one kilometre high into the atmosphere. On the ground, the reactor building at Unit 4 had transformed into an inferno in less than a minute, with at least thirty fires breaking out and its blazing core exposed.

In the immediate explosion, all of the noble gases inventory, xenon and krypton, in the core were blown into the atmosphere, mixed with an estimated six tonnes of disintegrated fuel in aerosols and particles. Over the next ten days as the graphite fire burned and for up to a month thereafter, the stricken reactor spewed a vast range of radionuclides into the environment. About 60 per cent of its core inventory of the short-lived isotope, iodine 131, was released. Isotopes of caesium and tellurium were emitted in large quantities. In all, Chernobyl released between 40 to 80 million curies of radiation; about 1,000 times more than the Windscale fire.

Only 20 km away from the reactor site, the neighbouring state of Belarus took the brunt of the fallout. On 27 April, workers entering a nuclear power plant in Sweden set off radiation alarms. The following day the International Atomic Energy Agency in Vienna received a telex from the Soviet government acknowledging that a major nuclear accident had occurred.

* * *

As the Chernobyl plumes spread, first in a north-westerly direction over Russia, Poland and Scandinavia, crossing over Belgium, Holland and into Britain and southwards over Central Europe and parts of the northern Mediterranean, the deposition pattern varied, largely depending on weather conditions. As did the political and public response to the crisis : the only common denominators in most affected countries were media hysteria, public confusion and the degree to which state agencies and authorities were found wanting in their response. Nuclear power may already have been unpopular in most Western European states, but what most characterised the reaction to

Chernobyl was the speed with which public anger turned either on their own nuclear installations or those in directly neighbouring states. As a French industry representative commented to *Newsweek*:

> Everywhere in Europe people turned their worry – and anger at being left in the dark – not against the author of the accident, but against their own governments and their own nuclear programmes.[2]

There were no demonstrations outside Soviet embassies in European capitals to protest against the delay in declaring the accident, but within days 30,000 people had marched on a nuclear power station west of Hamburg, demanding its immediate closure. The Danes were expressing concerns over the Swedish nuclear plant at nearby Barsebak, about 20 km from Copenhagen. In Austria, the government, which a few weeks earlier had contemplated a second referendum on its own mothballed nuclear power plant at Zwentendorf, decided to dismantle it. In Yugoslavia and Austria pregnant women were advised to 'remain indoors', although it was claimed that radioactive levels were 'not dangerous'.[3]

One hundred thousand people marched against nuclear power in Rome; and in Sweden, political pressure mounted for a complete phase out of nuclear power by a new deadline of 2010. French officials played down any health risk from contamination of its southeast region and claimed the Chernobyl cloud would be blown back over Russia within a week. In its first editorial on the disaster, the *Irish Times* noted: 'Irish people need no reminding of the danger from Sellafield on our own doorstep.'[4] The paper later, with stunning inaccuracy, reported the fire in the Chernobyl reactor as 'almost identical to that at Sellafield, then Windscale, in 1957'.[5]

Fallout in Britain was first detected by UKAEA monitors at their nuclear plant in Dorset in England. Radiation monitors at Sellafield also detected the contamination, when the levels of radioactivity coming down in heavy rainfall set off alarms at the site. But BNFL's corporate centre at Risley refused to allow any press statement that would link Sellafield and Chernobyl together in the public mind, especially since the site had only recently been in the news again for a spate of unplanned releases of its own.[6]

In Dublin, the meteorological service advised the NEB that the plume spreading over Britain would not reach Ireland. As late as 1 May, the Tanaiste and Minister for Energy, Dick Spring, reassured senators that Chernobyl radiation was 'unlikely to come near

Ireland'.[7] But the radiation cloud that brushed the south-east of France, Italy and the Iberian peninsula was carried in over Ireland and up the eastern half of the country by a weather front on Friday, 2 May. Throughout that Friday afternoon the Met Office tried in vain to get through to an agency swamped with queries from the public on phone lines that mainly registered busy tones all week.

'While the NEB and the newspapers were watching the South East England cloud throughout Saturday, radioactive rain was falling on Ireland,' the *Sunday Press* later reported.[8]

The inclement weekend weather left John Cunningham, Head of Radiation Monitoring at the NEB, feeling uneasy. By midday on Sunday, Cunningham had arranged for the collection of milk samples to check for iodine 131. The physics laboratory at University College Dublin, run by Dr Peter Mitchell, had more sophisticated analysis equipment than the NEB's own facilities, housed in St Luke's cancer hospital in Rathgar. On Monday afternoon, the UCD laboratory reported readings of 440 becquerels of iodine 131 per litre in certain milk samples, levels which, if continued, would be a cause of concern. In later samples the levels of contamination dropped steadily away. Air samples taken in Dublin between Friday and Sunday had shown radioactive contamination levels about twice the normal average. Such an increase, in terms of human health and safety, was still relatively minuscule and about half the levels being reported in parts of Britain.[9]

The Irish public was outraged to awaken on Monday morning to the news that the contamination unlikely to reach Ireland had in fact arrived, practically unnoticed, over the weekend. Public confidence in the NEB, and in any assurances offered by the authorities or the government, was compromised. Outrage was compounded by the delay in producing data on the extent and severity of contamination and by revelations of the NEB's monitoring inadequacies, lack of basic resources, such as an adequate number of telephone lines, and the reality that to all intents and purposes no civil nuclear emergency plan existed. Later that week in the Dáil, Charles Haughey reflected genuine public anger when he scathingly dismissed the NEB as 'little better than a public relations front for the nuclear industry'.[10]

But the Fianna Fáil leader was playing a wider political game, far beyond immediate public concerns about Chernobyl fallout, the competence of the national nuclear watchdog or the government to cope with a nuclear crisis. Or indeed about the genuine public fears about Sellafield radiation. On 6 May 1986, for the second time in the space of three months, Fianna Fáil tabled a motion for debate in the

Dáil calling for the closure of Sellafield.[11] As far as Haughey was concerned, Sellafield was an ideal political stalking horse 'to test this new friendship and co-operation which is supposed to exist between the two Governments'.[12]

* * *

'Frigid' was the best description Dr Garret FitzGerald could offer of the state of Anglo-Irish relations when his government took office in November 1982. Mrs Thatcher was reputed to have personally considered Charles Haughey charming and urbane but their respective political mindsets, particularly on solutions to the problem of Northern Ireland, were irreconcilably different. Dr FitzGerald's experience as Foreign Affairs Minister in the 1970s had left him well acquainted with the finer points of the British establishment's sensitivities on Northern Ireland and the Republic's role in relation to it. Dogged persistence and seemingly inexhaustible patience were prerequisites to secure any progress.

Within three years such persistence paid off. A diplomatic triumph, the Anglo-Irish Agreement of 1985 laid an important foundation stone for the future peace process. The agreement, formally signed in December that year, constituted a new departure in the relationship between Ireland and Britain in that, for the first time, the Republic's right to an input into how Northern Ireland was governed was explicitly acknowledged by international treaty. Inevitably, it was the sort of agreement that held something for everybody directly involved but that nobody much liked, since it demanded flexibility from all parties on previously cherished principles, including the Republic's constitutional claim to jurisdiction over all of Ireland. FitzGerald was dismayed, but hardly surprised, when the Fianna Fáil leader refused to support the agreement. Moreover, Haughey appeared cynically determined to undermine it at every turn.

Haughey's party strategists perceived Garret FitzGerald as 'soft' on Sellafield and were increasingly convinced, especially after the Anglo-Irish Agreement, that FitzGerald was reluctant to confront Margaret Thatcher or her government on the issue. Thereafter, Haughey, described by Martin Mansergh as 'one of the first environmentalists', never missed an opportunity to highlight Britain's nuclear perfidy, in tones laced with strong nationalist sentiment.

Haughey had branded the 1984 Black Report 'a dreadful piece of whitewash' and demanded imprisonment of those responsible for its 'lies'. Two years later, he felt vindicated when it emerged that the true scale of the pre-1957 emissions from the Windscale stacks was not 440 gm of uranium oxide as originally represented to Black, but more in the region of 20 kgs.[13]

A House of Commons Environment Committee Report on Sellafield discharges, published in February 1986, added further grist to the Fianna Fáil mill. Entirely oblivious to its likely impact in Ireland, the House of Commons report noted that Sellafield was 'the largest dispenser of nuclear waste into the environment in the world'. Newspaper headlines that the Irish Sea was the most radioactive sea in the world coined a mantra for endless repetition by Irish anti-Sellafield campaigners for the next twenty years. Advice from the NEB that its monitoring proved radiation doses to Irish people from Sellafield discharges were negligible and represented no significant health hazard was always included in official statements, for the record, but then usually summarily discounted both by politicians and the media.

* * *

'This is an international rather than a bilateral issue and ... it should be resolved under the provisions of the Euratom Treaty,' Dick Spring summarised the coalition government's Sellafield policy. In his capacity as Minister for Energy, Spring vigorously pursued the concept of a European nuclear health and safety inspection force, that would operate independently of national regulatory bodies, with the European Commissioner, Stanley Clinton Davis, and at Environment and Energy Council meetings.

However sympathetic the European Commissioner may have been to the Irish proposal, especially after Chernobyl, there was never any realistic prospect that the nuclear states in Europe would agree to any such development under the Euratom Treaty. If Britain didn't veto the proposal, France surely would. Irish politicians either refused to recognise this political fact or chose to ignore it. Concerns about Sellafield discharges raised at the broader international fora of IAEA meetings and at the Paris Commission were hardly any more effective, but at least had a propaganda value in national terms.

Following the beach incident, Spring met his counterpart in the British government, Patrick Jenkin, in February 1984, the first ever bilateral meeting of a British and Irish minister over Sellafield. The Irish Labour leader received assurances there would be no repetition of the incident. An Irish–UK Contact Group of officials and diplomats was set up, meeting at six-monthly intervals. A new notification procedure for nuclear accidents was also agreed, supplementary to the notification arrangements already in place between the NEB and the UK Nuclear Installations Inspectorate.

In the year up to November 1986, Spring's department was notified of nine separate incidents at Sellafield. While none was particularly serious, nor held any radiological implications for Ireland, the list included a series of radioactive leaks that contaminated Sellafield workers, the discharge of unmonitored effluent to the Irish Sea and incorrect labelling of materials sent to Dounreay that might, conceivably, have resulted in a criticality incident. It was the cumulative effect of so many incidents that undermined confidence in the management of Sellafield, Spring said.

The government had responded to the health issues raised by the Black report and the Sheehan study on the Dundalk Down's syndrome cluster by initiating a study on childhood leukaemia along Ireland's east coast. In November 1984, the Department of Health engaged the Department of Community Medicine and Epidemiology in University College Dublin to form a group to conduct the study and a parallel examination of incidence of Down's syndrome throughout Ireland. The Northern Ireland government also established a leukaemia and Down's syndrome study, led by the epidemiologist Dr Sidney Lowry.

When it reported in late 1986, the expert group concluded there was no evidence of an excess of childhood leukaemia cases on Ireland's east coast and nothing to link incidence of the disease in the Republic to Sellafield radiation. The Northern Ireland study reached similar conclusions. The UCD group had run into a brick wall on data collection prior to 1977, however, because at the time Ireland, almost uniquely among modern European countries, had no national cancer registry. As well as further research, establishment of a national cancer registry formed a key recommendation.

Garret FitzGerald regarded the risk to Ireland from Sellafield as a genuine public and political concern but preferred to leave implementation of his government's international strategy on Sellafield to his Minister for Energy, Dick Spring. FitzGerald delayed making direct representations to Margaret Thatcher, seeking a review of Sellafield discharges to the Irish Sea, until a meeting between the two prime ministers on 19 February 1986. A month later the British Prime Minister wrote to the Taoiseach providing assurances that Sellafield discharges were within authorised limits. The letter also pointed out that, once again, Britain's regulators were examining safety policy at Sellafield.

At national level, Dr FitzGerald was unfazed by Haughey's relentless campaign to make Sellafield a core bi-lateral issue between the two governments.

'The priority was the agreement in Northern Ireland. There was no doubt about that and Mr. Haughey was opposed to the agreement. He could push [Sellafield] and hope to get some political kudos from it. When you're in opposition you can demand all sorts of things in the clear knowledge that they don't have to go anywhere. His pursuit of the issue was designed to cause difficulties, but of course there's nothing unusual about that in politics,' Dr FitzGerald said.

FitzGerald, whose essential strain of political pragmatism was often lost sight of by his contemporaries, himself failed to see the point in making demands of the British government that stood no chance of being conceded and would serve only to irritate a functioning, if fragile, relationship.

The frequent taunts and jibes that FitzGerald's party was afraid to broach Sellafield with the British, that they were 'cuddling up' to the British, or the barely concealed xenophobia that decried the 'traditional, characteristic attitude of Fine Gael governments towards our neighbours ... slow to rock the British boat'[14] cut no ice with the then Taoiseach. Fianna Fáil's efforts to use Sellafield to disrupt Anglo-Irish relations were entirely futile.

'It [Sellafield] was never allowed to get in the way because of the way in which we dealt with it,' FitzGerald later recalled.

* * *

Fianna Fáil and its leadership had been engaged in a cuddling up of sorts of their own, with Greenpeace International. Throughout the mid-1980s, several high-level meetings took place between Greenpeace representatives visiting Ireland and Fianna Fáil. Haughey paid a courtesy call on a Greenpeace ship, bound on yet another Sellafield mission, when it docked in Dublin Port.

The interests of Ireland's largest political party in demanding all-party commitment to a policy of closing down Sellafield matched those of the international environmental protest group on at least that one point. Greenpeace proved helpful in other respects. The organisation supplied background information and advice on policy initiatives utilised copiously by Fianna Fáil deputies in the numerous parliamentary challenges to the coalition government's stance, some of which, such as Ireland taking international legal action against the UK over Sellafield discharges, ultimately became central to Ireland's campaign.[15]

It was third time lucky for Fianna Fáil when it tabled another motion in the Dáil calling for the closure of Sellafield on 3 December 1986. The motion was prompted by Dick Spring's description of Sellafield as a 'dangerous monstrosity' and his call, in the previous week, for an international campaign to close the site. What Fianna Fáil gleefully presented as a damascene conversion to their cause, Spring explained as the logical culmination of his own and his government's frustration.

'The Government have refrained up to now from calling for closure of a plant located in another sovereign state,' he told the Dáil. 'Instead we wished to be assured that the plant could operate safely without posing any threat either to our population or to our environment. We in the Irish Government took every reasonable measure to encourage the establishment of an independent inspection force which could assure us that the plant was capable of operating safely and without any hazard to this country.'[16]

The independent nuclear inspectorate proposal was bogged down at European level. In the UK itself, the regulator had published yet another predictably damning report questioning the competence of Sellafield management and calling for more investment in safety at the site. But the coalition's ready acceptance of a Fianna Fáil motion that it had previously voted down twice in the same year also had less to do with Dick Spring finally seeing the light than the government's own immediate political priorities and the imminence of a general election.

After four years in office, the government barely commanded a parliamentary majority any longer, with backbenchers in both parties determined to go their own way on various issues. Both FitzGerald and Spring recognised the impossibility of agreeing a budget for the coming year. Fine Gael was committed to the introduction of a programme of dramatic cuts in public expenditure to salvage an economy on the verge of bankruptcy. Labour in government baulked at the implications for the disadvantaged in Irish society.

Rather than break up the administration that December and, as Dr FitzGerald put it, 'spoil everyone's Christmas', the leaders made a gentleman's agreement to stage a civilised parting of the ways early in the New Year. The government was also committed, as part of the Anglo-Irish Agreement, to the passage of a Bill to sign the 1977 European Convention on Extradition and thereby establish, for the first time in the history of the Irish state, extradition arrangements with Britain that no longer excluded offences claimed as 'politically' motivated. Accepting a Fianna Fáil motion to close down Sellafield to secure enough of their own backbench votes to support the Bill, was a small political price to pay.

The Extradition (European Convention on the Suppression of Terrorism) Bill, 1986 was introduced to the Dáil on 4 December. Fianna Fáil, with political astuteness, accepted the principle of the Bill, but tabled amendments that would constrain the smooth extradition process under the convention. To see it through its critical amendment stages and in the final vote for passage of the Bill, the government was obliged to rely on the casting vote of the Dáil Chairman, the Ceann Comhairle. In his last meeting as Taoiseach with the UK Prime Minister later that month, Dr FitzGerald conveyed the adoption of the resolution, supported by all parties in Dáil Éireann, demanding the closure of Sellafield.

* * *

Greenpeace opened its Irish office in Dublin in April 1987, a month after Charles Haughey and Fianna Fáil returned to office in a minority administration.

'The office was basically set up as a short to medium term affair to deal with the Sellafield issue,' according to John Bowler, who had been appointed head of Greenpeace Ireland, 'which I think was a good strategic move in that Ireland was the closest neighbour to the UK ... Greenpeace thought that they could do a lot politically here, which happened to be the case.'

Bowler was tailor-made for the role as Ireland's Mr Greenpeace. He grew up in Dublin's north side, in Malahide, the son of showbusiness parents who had worked as entertainers to the Allied forces at the end of the Second World War and bore personal witness to the ravages of the war and the concentration camps in Europe. John's parents were vegetarians and active in CND in the 1960s. Growing up, he was strongly influenced by his father's ideals: 'He was someone who was involved in all levels of what I would call justice; justice for people, the worker; justice for animals and for all who inhabit this world.'

Bowler took leave of absence in 1981 from his job as a typesetter in a printing works to campaign against seal culls in the west of Ireland. Together with some friends he formed an Irish branch of Sea Shepherd, the breakaway group from Greenpeace. Moving on to the nuclear issue, in September 1986 Bowler organised a round table conference in the Grand Hotel, Malahide, entitled 'A Common Irish Nuclear Policy'. Representatives of all parties with an established parliamentary presence at the time were invited, and attended, along with most of the north Dublin sitting deputies, Irish CND and Earthwatch, the Irish offshoot of Friends of the Earth.

'We actually came out with more or less an agreed position on a lot of things,' Bowler said. The success of the round table conference was, he believed, instrumental in his selection for the Greenpeace Ireland job.

One of Greenpeace Ireland's first meetings was with the new Taoiseach, Charles Haughey, early in May 1987. 'When he was in opposition he was very supportive of Greenpeace, I believe,' John Bowler said. 'I suppose as in most cases, [as Taoiseach] he was a little more reticent to be open with us and openly supportive of us.'

Bowler recalls mostly dealing with Ray Burke, who held the Energy portfolio in the new government. The linchpin of Fianna Fáil's parliamentary campaign to position Sellafield as a core bilateral issue between Ireland and Britain, Burke never lost an opportunity in the media, parliament and on the doorsteps of his constituency in north county Dublin to press for the closure of Sellafield. All means would be used, he said, including legal action against Britain in the international courts. His party's 1987 election manifesto also included a commitment to replace the NEB with a new body, the Radiological Protection Institute of Ireland, with a wider brief for radiological protection of the Irish people.

In his first weeks as Ireland's new Energy Minister, Burke wrote to his counterpart in the UK, Peter Walker, demanding the closure of Sellafield. Greenpeace Ireland lobbied Burke to press a motion calling for the closure of Sellafield at the Paris Commission officials' meeting, due to take place in Cardiff in June. Burke was devising a strategy for an alliance with the governments of Nordic countries, another tactic strongly favoured by Greenpeace, whose international lobbying strategy was to move several countries towards a single common point of view on an issue at the right point in time. Fianna Fáil in government also maintained their predecessor's policy of seeking the establishment of an independent nuclear inspection force under the Euratom Treaty.

In the Dáil on 3 June 1987, Minister for Environment Padraig Flynn reported the progress of Ireland's 'shut Sellafield' motion at the Cardiff meeting. 'Unfortunately, it did not receive widespread support from the participating countries,' he announced. In fact, nine of the ten delegations at the Commission meeting had voted against it although the minister could claim some headway might be achieved on a more general Irish motion calling for the use of best available technology to reduce radioactive discharges to the marine environment.[17]

* * *

In an action timed to coincide with the Paris Commission meeting in Cardiff, Greenpeace divers had finally succeeded in blocking the Sellafield pipeline on 3 June 1987. BNFL treaded a path of at least twenty court actions across Europe to finally obtain an injunction in London on 1 June against the Greenpeace action.[18] Tensions ran high between Greenpeace and BNFL divers at the scene and, at times, threatened to turn violent. The pipeline block provided a media spectacular for Greenpeace, and for John Bowler, back in Dublin. 'We were of course trying to push the Irish media to keep up the pressure on the Irish delegation to get agreement on the proposal,' he said.

To commemorate the thirtieth anniversary of the Windscale fire, Bowler held a press conference in Dublin and invited an all-party parliamentary delegation to a protest at the gates of Sellafield on 10 October 1987. Once again, Greenpeace Ireland was gratified by extensive Irish media coverage, although one protesting parliamentarian expressed disappointment that only 'about 250 people turned up and I expected thousands would turn up'.[19]

Irish policy was being pulled in two opposite directions by politics and science respectively. The science, as enunciated by the NEB, sought to place Sellafield discharges and nuclear accidents within context, and one which representatives of the government were obliged to take into account; the politics demanded the opposite. A clash was imminent and inevitable.

17

Battle Stations

'Clandestine and willing to risk thousands of lives on both sides of the Irish Sea.'
           Greenpeace comment on proposed Trawsfynydd reactor test

Post Chernobyl, an announcement that put 'experiment' and 'nuclear power plant' in the same sentence was enough to raise the hackles not just of any self-respecting local community but every environmental group worthy of the name. Inevitably, it would draw the attention of the national and international media.

The 500 MW Trawsfynydd nuclear power plant, a twenty-storey monument to the functional ugliness of 1950s industrial architecture, incongruously located in the spectacularly scenic Snowdonia National Park in North Wales, was the only nuclear power station in Britain ever built inland. Trawsfynydd came on line in 1965. The station drew its water supply from an artificial lake, constructed as part of a hydroelectric power scheme in the 1920s.[1]

The Trawsfynydd plant was ill-fated. In February 1986 a faulty valve had caused the release of fifteen tonnes of its carbon dioxide coolant to the environment. The NII judged the incident of little radiological consequence, posing no health or safety threat to the local population, but there were few locally, and among the Irish body politic, who would have agreed with that assessment.

Frank Turvey was the NEB's resident expert on nuclear reactors. When the Central Electricity Generating Board in Wales announced a test on its magnox station at Trawsfynydd, scheduled for mid-February 1988, Turvey requested the specifications from his counterparts in the UK Nuclear Installations Inspectorate.

The Irish media picked up the Trawsfynydd story from British newspapers over the weekend of 16 January 1988, reporting that the EC Commissioner, Stanley Clinton Davis, had written to the British

Energy Secretary to point out that in the wake of Chernobyl all 'dangerous experiments' required commission notification. The proposed experiment was denounced by the Taoiseach, Charles Haughey. Labour leader Dick Spring described it as a 'hostile act' by Britain against Ireland. Ray Burke was in the US on a promotional trip for the ESB. Nobody, including the department, had as yet approached the NEB for their views.

During Monday, January 18, Frank Turvey fielded calls from Irish journalists. The test, Turvey explained, bore no resemblance to the experiment that had exploded the Chernobyl reactor. It was a standard test routinely performed on all reactors to see if, in the event all electrical power was withdrawn, the reactor would lose its residual nuclear heat by means of natural convection. Turvey was surprised this test had never been carried out on the Trawsfynydd reactor, without which safe operation of the station could not be guaranteed. In his view, if the test failed it would simply reinforce arguments for permanent closure of the power station.

In a call from the department later that day, Turvey said he hadn't time to provide a written note on the Trawsfynydd test. He sought advice on how to respond to any media questions about briefing the minister. Since there was no opportunity to advise the minister before the issue had emerged in the newspapers, the department official suggested, it would be fair to say Burke would be immediately briefed on his return from the US.

As the minister flew into Dublin next morning, Frank Turvey was being interviewed on Ireland's premier radio news programme, 'Morning Ireland'. Asked why the NEB welcomed the test and if the Taoiseach, his own minister and the Labour Party leader were all ill-informed in their condemnations, Turvey replied: 'You cannot bend the truth for political reasons.'

'The truth is the truth,' he said. 'This is the Board's technical judgement and if we are asked what it is we are bound to give it, straight from the hip, so to speak.'[2]

'I came out with guns blazing,' Turvey admitted years later. 'It was carefully considered and I was just waiting for the question, to give the answer.'

* * *

He had been put in an 'intolerable position', Burke told the chairman of the NEB, Dr George Duffy and chief executive, Noel Nowlan, later that afternoon. Both he and the Taoiseach had been 'extremely embarrassed' by Turvey's remarks. He could not understand how the NEB could intrude 'into the political field in such a matter'. He demanded to know 'what the Board intend to do about Mr Turvey'.

Before meeting the NEB, Burke had attended a cabinet meeting. He had already sent a letter to the British Energy Secretary, Cecil Parkinson, seeking an immediate meeting and laying out in no uncertain terms his, and the Irish people's, opposition to the test. Later he phoned Commissioner Clinton Davis to put the Irish government's protest on the record in Europe.

Desperate to stamp out the fuse Turvey had lit under their own relationship with the government that morning, the NEB board had released a statement at noon, distancing itself from his remarks. The NEB had no intention of prejudging the NII or the European Commission's conclusions on the safety aspects of the Trawsfynydd test, the statement said. The minister and his departmental secretary, J.C. Holloway, now insisted they publish another, taking specific issue with Frank Turvey's statements on 'Morning Ireland' and endorsing the minister's line with the UK Energy Secretary.

Holloway further suggested the NEB's sole function was to advise the minister and it had no business talking to the press. George Duffy retorted that great harm had been done in the post-Chernobyl period by 'alarmist misinformation' and in Europe alone an estimated 40,000 abortions had been carried out as a result. The statements of characters such as Mr John Large, the NEB chairman said, were much to be deplored. The minister stated that since nuclear power was a highly political issue, in future any NEB statements should be cleared in advance with the department.[3]

The government had decided to abolish the NEB and replace it with a new radiological protection institute, Ray Burke informed senators in a special Seanad debate the following day.

'Whatever attitudes may have manifested themselves in the past in the Nuclear Energy Board, a pro-nuclear approach will in no way be reflected in the charter of the national radiological protection institute,' Burke assured angry senators. The new institute's terms of reference would 'accord with the present climate of opinion in Ireland surrounding the nuclear industry and the realities of the world in which we live,' Burke promised.[4]

As Turvey waited to learn his fate from an NEB emergency board meeting, he could take some consolation from the fact that he was not entirely bereft of support.

The scientific staff of the NEB wrote to the board: 'We support the view that the "truth should not be bent for political reasons" and regret that the present policy of the Board does not reflect this.'[5]

Dr Ian McAulay, then head of the environmental radioactivity laboratory in TCD's Department of Pure and Applied Physics, was among those independent scientists who expressed disappointment at the board's failure to support Turvey's statements.

'If the [NEB] is not prepared to give full backing to the technical judgement of its own experts ... it can hardly expect to have the confidence or support of scientists outside its own staff in the future,' McAulay warned in a letter to the chairman.

'It will be assumed, apparently with some justification, that statements from the Board or its staff will reflect only the views of whatever Minister is in office at the time.'[6]

In a second wave of the story the media alleged Ray Burke had 'gagged' the NEB, and the so-called independent nuclear watchdog was no more than a puppet on a string in the minister's grasp. Burke retaliated that the sole function of the NEB was to advise the minister on matters relating to nuclear energy, 'not to publicly comment or express opinions on the content of such advice'.[7]

Frank Turvey felt the row had helped draw a clear line between the role of an independent regulator and the political imperative. But at the time he was shocked that: 'So many people, some of whom I respected too, said to me "That was a bad mistake Frank. You made a gaffe."'

Ray Burke, for his part, was unhappy with the way his British counterpart dismissed his objections to the Trawsfynydd test and the demand to close Sellafield, now joined by demands to shut down all the magnox power stations and permit no further expansion of Britain's nuclear industry. Any proposals for long-term nuclear waste management that NIREX, the body established by the UK government in 1982 to develop solutions to the mounting nuclear waste problem, was putting forward, were also unacceptable.

'I am not satisfied with the response I have been getting from the British on the legitimate fears we have as a nation,' Burke complained.

'We have used every diplomatic channel available to us in the UK and we are considering the situation of going a step further and taking legal action because we feel that you can only depend on the diplomatic process for so long.'[8]

The advice of the Attorney General on Ireland's legal options was awaited, he said.

* * *

At the beginning of February 1988, the CEGB announced it was postponing the Trawsfynydd test.

'We have received many representations from local people and we are impressed with the sincerity of their concerns,'[9] the CEGB said.

The test never took place. In Ireland, its cancellation was hailed as a victory for the government's vigorous lobbying, as was the premature shut-down of the station a few years later in 1993.

The entire episode displayed many of the elements of a comic farce. But it was of greater significance than a head-to head contest between two strong-willed and single-minded individuals, Burke and Turvey.

Turvey, as the NEB technical expert, felt he had a public duty to present the facts about the Trawsfynydd test fairly and squarely. On his own admission he had also chosen that moment to confront the prevailing political orthodoxy. Burke, like all politicians, could only portray certainties or face ridicule and accusations of equivocation, in the media, from the opposition, his own party and his own electorate. His boss, Taoiseach Charles Haughey, had set the political bottom line by condemning the test. Any challenge to that line, particularly from within the public service, was a challenge to the authority of the government itself, and Burke, the consummate politician, could never have conceived of it otherwise.

Opposition to Britain's nuclear industry was, Burke declared, 'the one issue that unites the whole nation'.

The release of British government papers on the Windscale fire on 1 January 1988, under the thirty-year rule, had revealed the extent to which the Penney Report into the fire was censored before publication. If they hadn't been before, Irish politicians were now convinced that the British government was as cavalier about public health and safety as the nuclear industry they sponsored. That the main object of the 1957 so-called cover-up was not to deceive anyone except the Americans, as part of Britain's foreign policy objective to secure a 'special relationship' with the US on nuclear weapons, never occurred to them.[10] At every level and within every party, Irish politicians became inspired with a messianic zeal to protect the British public, as well as their own, from Perfidious Albion's nuclear ambitions.

When the Bill to establish the Radiological Protection Institute of Ireland was finally brought forward in 1990, long after Ray Burke had moved on from Energy, explicit provision was made for the institute to take directions from the minister and be held directly accountable to him. Fine Gael's Richard Bruton opposed even an initial reading of the Bill.

'If the British Government moved legislation in the Houses of Parliament tomorrow suggesting that their National Radiological Protection Board which is answerable directly to Parliament and not to Government at all should be answerable to the Secretary for Energy, as is implied in this Bill, we on this side of the Irish Sea would be absolutely incensed,' he said.

But the reasoning behind Bruton's opposition to the Bill revealed his own growing anti-nuclear paranoia. 'It is very hard for the public to believe,' he added 'that the Department [of Energy] have so completely changed their spots that they are no longer considering nuclear capacity.'[11]

* * *

'All through the years that we were working here and had an office here, Sellafield was the number one issue that the public responded to, the number one issue that the public contacted us about; I would say the number one issue for the media when they wanted to deal with issues like this,' John Bowler recalled.

'And let's be honest, it was a wonderful issue for politicians of every colour and flavour, because they didn't have to do much, let's put it that way, because it wasn't our problem.'

Fianna Fáil had bought into the Greenpeace International agenda to close Sellafield in every aspect including international court action. The problem for Fianna Fáil and Ray Burke in government was that Greenpeace did not discriminate between political allegiances. From a Greenpeace perspective, Sellafield was also a wonderful issue, as John Bowler put it, 'especially for opposition politicians to hammer the minister'.

Bowler had politicians on his books, so to speak, from both government and opposition parties right across the political spectrum. Key opposition players included Fine Gael's Richard Bruton and his brother, John, the Workers Party Deputy, Pat McCartan, Pat O'Malley of the new liberal Progressive Democrats group, and most members of the Labour Party's front bench. All could be regularly relied upon to harass Burke at every turn, especially on the promise of international court action.

On this Burke and his successor in the department, Michael Smith, hedged their bets in their parliamentary replies, not yet prepared to admit publicly that the Attorney General's advice to government was that any such legal action was likely to fail. Burke set the tone for the standard parliamentary reply some nine months after he had formally referred the court action question to the AG's office in the wake of the Trawsfynydd row.

'It would be wrong, and a very serious setback to us, if we took a case which proved to be unsuccessful,' Burke told the Dáil in October 1988. Fine Gael might do that sort of thing in government, he claimed, 'but Fianna Fáil will not do it'.[12]

On issues closer to home, a national nuclear emergency plan couldn't be published, Burke maintained throughout 1988 and early 1989, because it needed to be properly tested and monitoring equipment and computers had yet to be bought for the NEB. The question of a national repository for Irish nuclear waste was dealt with in classic political style – firmly placed on the long finger. Michael Smith was one of the series of Ministers for Energy whose proposals to deal with this necessary piece of infrastructure, which would be undeniably unpopular and difficult to sell to any prospective host community, were queried.

'The present arrangements for disposal of radioactive waste arising in Ireland cannot continue indefinitely,' Smith stated. 'We all acknowledge that some positive action is needed but the question of identifying a site has presented some difficulties.'[13]

Meanwhile, the Irish government objected vociferously to any proposals in Britain to deal with the mounting issue of nuclear waste, reacting angrily to media reports of site investigations anywhere in the UK, in particular any prospect of a national repository at Sellafield, where most of Britain's medium-level nuclear wastes were concentrated.

The government also refused to sign up to any of the international nuclear accident liability conventions, which would protect the rights of Irish citizens in the event of a nuclear accident in Britain that impacted on Ireland, on the grounds that the conventions were not in the national interest.[14]

A policy that relied on the moral force of some day securing non-binding resolutions at the Paris Commission or what the British embassy officials routinely dubbed as nuclear megaphone diplomacy was not so much, as PD Deputy Pat O'Malley described it, 'a bit of a joke'. In truth, it was more a shambles, but one that in the late 1980s played well with a credulous media and the general public.

* * *

Charles Haughey's government enjoyed no such latitude when it came to other pressing problems of governance. Unlike De Valera in the 1930s, who had sought to dismantle the Anglo-Irish Treaty, Haughey in government assiduously worked the Anglo-Irish Agreement he had so bitterly opposed. Haughey rationalised his position on the grounds that no Irish government could unilaterally abandon an international treaty.

Albeit with some modifications, his government also upheld the controversial 1986 Extradition Act. But there were no special conferences with the British Prime Minister to progress co-operation on Northern Ireland, as Garret FitzGerald had envisaged when the agreement was first concluded. Haughey's discussions with Thatcher were restricted to fringe meetings at European Councils. Haughey was paranoid about the hostility towards him in the British establishment and at one point is said to have believed British agents were conspiring to bribe members of his party to oust him from leadership.[15]

The then Taoiseach had secretly, and at some political risk to himself, embarked on a course of his own to bring about reconciliation in Northern Ireland. This clandestine departure involved making third-party contacts with Sinn Féin and the Provisional IRA with a view to promoting a long-term, more inclusive, settlement of Northern Ireland's political affairs, another stepping stone on the long path towards the Good Friday Agreement several years later.

Nor did Haughey flinch when it came to radical surgery on Ireland's ailing economy. Haughey secured Fine Gael's parliamentary support for swingeing economic cuts to restore balance to the public finances. His Minister for Labour, Bertie Ahern, was charged with the task of securing the support of trade unions and employer bodies for the agenda of economic reform, through a process of social partnership and national agreements on wage policy.

Haughey also presided over the last single-party Fianna Fáil administration in twentieth-century Irish politics. From its foundation, Fianna Fáil had rejected coalition governments on a point of principle. Charles Haughey's divisive leadership had seen off many of the Fianna Fáil old guard by the late 1980s, including Des O'Malley, finally driven from Fianna Fáil for alleged 'conduct unbecoming' in 1985. O'Malley's new liberal party, the Progressive Democrats, a niche party that attracted widespread support among Ireland's socially liberal but economically conservative middle classes and the business community, was soon mopping up votes that might traditionally have found a natural home in Fine Gael and Fianna Fáil. Following a snap election in 1989, a political gambit for an overall parliamentary majority that failed, Haughey was left with no option but to enter a coalition government with his old nemesis, O'Malley, and the Progressive Democrats party, in order to stay in power.

While his personal life and style exhibited all the characteristics of a corrupt sub-Saharan potentate, Haughey's political persona terrified his political friends as much as his enemies, and generally brought out the worst in everyone caught at close quarters with him. Nevertheless, this enigmatic, mercurial and undoubtedly paranoid man was a figure of considerable political achievement, including setting the course for Ireland's economic transformation in the 1990s.

'He did far more than his critics ever did,' another former Taoiseach, Liam T. Cosgrave, would say of him when Haughey was finally forced to resign in January 1992.[16] That he accumulated a vast store of personal wealth, built on the generosity of business friends and acquaintances, and financed a lavish lifestyle sometimes courtesy of his own party's funds, would later be counted among his less admirable achievements.[17] Haughey was also, undeniably, the political architect of an anti-nuclear mindset that would persist long after his time.

* * *

Margaret Thatcher had been elected to power in 1979 on a promise that trade unionists – and especially the powerful National Union of Mineworkers – would no longer govern Britain by proxy. Thatcher was equally determined to reform the British energy market. She privately acknowledged that a day of reckoning was inevitable with the NUM and its leader, Arthur Scargill. The Prime Minister backed off confrontation with the miners in 1981, but by 1984 the government was satisfied it could withstand a prolonged strike by the coal miners' union. Enough reserve raw materials were stockpiled at the power stations and Britain's nuclear stations were primed to make up the energy shortfall.

The year-long miners' strike that began in 1984 inevitably increased pressure on waste management and fuel handling processes at Sellafield and undoubtedly precipitated some of the incidents and safety lapses recorded at the site in that period. It added to the long-term already burgeoning waste management problems at the site, particularly magnox fuel reprocessing. It also did little to improve already strained relations between the coal miners' union and the nuclear communities.

'Scargill was at the height of his popularity, at the height of his fame. He had lost the miners' strike, but he was revered by the left of the Labour Party,' Jim Innes, a veteran Labour Party activist, recalled.

At the Labour Party conference in Blackpool in 1986 Scargill led the charge against Britain's nuclear industry. Nuclear power, he declared, was unsafe, uneconomic and unnecessary. In a vote committing Labour to phase out Britain's nuclear industry the trade union bloc conference vote was pivotal to Scargill's victory.

Earlier that year, shop stewards at Sellafield had got wind of the conference motion.

Sellafield, with its 7,000-strong unionised workforce and some 4,000 contract workers, most of whom were also union members, had its own elite cadre of elected full-time shop stewards, equipped with their own offices and facilities paid for by the company. The main general workers' union on the site was the GMB. On some other BNFL sites the T&G was the dominant union. BNFL employees were also members of Amicus and UCATT and a variety of smaller unions.

The Sellafield shop stewards organised a conference in Carlisle in early spring 1986 bringing together their colleagues from the other BNFL sites of Capenhurst, Chapelcross, Springfields and Risley, out of which the National Campaign for the Nuclear Industry (NCNI) was born. The NCNI lobbied the 1986 Labour Party conference, standing en masse outside the hall, waving banners and distributing leaflets. They lost the vote.

As the NCNI saw it, they risked losing everything: the Labour Party was now formally against them and their trade unions were not up for the battle, so they had to do it themselves. For the Labour MP for Copeland, Jack Cunningham, facing a general election the following year, the stakes were equally high.

'It's not often that an MP goes into an election with 7,000 of his own constituents working in an industry that his party wants to shut down. It's a rather unusual position to find yourself in; unique I would have thought in the history of British politics,' Innes observed.

If the T&G – by now avowedly anti-nuclear – was turned around, then the Labour Trade Union bloc vote could be converted to a narrow majority in favour of nuclear. Changing the political climate within the rank and file of Labour itself was a more demanding task on which playing for time was the best immediate option.

'Cunningham and the shop stewards would have reckoned that the only way to win a battle within the Labour movement was to fight it by the rules of the Labour movement,' Innes recalled. 'Later on they began to understand the power that came from the support of their communities, but in the early days the NCNI played the only card they trusted, the union card.'

In early 1987, Innes, who was well known to the local MP, Jack Cunningham, through his work for the Labour Party at national level and in by-elections throughout the country, was asked by Harold Bolter if he would use his campaigning expertise to help the NCNI. The company would pay his fee and expenses.

Bolter was taking a risk. There was no guarantee the NCNI would accept Innes, even if they obviously needed professional campaigning advice. As it turned out, the shop stewards took the pragmatic view that, far from compromising their integrity, this particular contribution to their campaign was just part of the services they were due from the company.

One of the NCNI's first projects was to find out just how big Britain's nuclear workforce really was. *Shut Them Down*, a Greenpeace campaign leaflet published in 1986, had 'totally' rejected what it called 'the frequent but utterly unsubstantiated claim that 140,000 jobs will be lost if nuclear power is abandoned'.

Weeks of hard graft, phoning all of Britain's civil nuclear sites and asking for the numbers of workers employed – information often refused by site management but relentlessly pursued through site unions – was eventually rewarded. The NCNI used a multiplier of 0.6 to estimate the number of indirect jobs supported in each nuclear community.

'Once the sums had been done, the NCNI's estimate in 1987 of direct and indirect workers was just under 170,000,' Innes said. 'The crucial fact was that no anti-nuclear NGO felt able any longer to challenge that figure.'

Labour's 1987 election manifesto – drafted by a shadow cabinet that included Copeland's Jack Cunningham – contained one remarkable paragraph: 'We share national concern about the problem of nuclear waste. We will ensure a safe future for Sellafield and develop a new strategy for the monitoring, storage and disposal of nuclear waste.'

Just a year after his party conference had voted to phase out the entire British nuclear industry, Cunningham secured a sop to West Cumbria's nuclear community in its election manifesto. His electoral margin remained tight, however, retaining the Copeland seat for Labour in 1987 by a majority of just seventy-nine votes.

* * *

Britain's Minister for Energy in the early 1980s, Nigel Lawson, believed that large monopolies resulted in inefficiency, higher prices for industry and domestic consumers of electricity and, particularly the case as far as the nuclear component of Britain's energy sector was concerned, a seemingly endless obsession with innovation and indulgence in pet projects. Lawson was determined to break up the CEGB and impose a price market for electricity supply where none had previously existed.

Following the defeat of the miners, there was complacency in government circles about Britain's energy security of supply. Britain had its own oil and gas, courtesy of the North Sea fields. UK electricity prices, both for industry and the consumer, compared favourably with those of Britain's main European competitors.

Cecil Parkinson introduced the market liberalisation experiment in 1990, privatising most of Britain's conventional electricity sources. Since the 1950s, nuclear energy had contributed markedly to stability and security of electricity supply in Britain.

Almost at the eleventh hour, nuclear was dropped from the privatisation package: the real costs of nuclear generation were unclear mainly because there was no plan in place, and no cost estimates, for long term disposal of nuclear waste. The old magnox stations posed a particular problem: reprocessing contracts between the CEGB and BNFL were labyrinthine in their complexity. The nuclear industry, including BNFL, the government decreed, could prepare for privatisation over the longer term. Meanwhile, the new stand-alone nuclear generating companies – Nuclear Electric in England and Wales and Scottish Nuclear – must prove themselves fit for competitive survival in the privatised marketplace.

Energy privatisation effectively put a brake on any further investment in new nuclear build. It also brought a halt to any further government investment in fast-breeder or other experimental reactors. As the national electricity supplier and distributor, the CEGB had been content with modest profit margins, never more than 5 per cent. The CEGB was also amenable to government intervention on preferred source of supply. In a competitive marketplace, profits of 10 per cent were the minimum acceptable to shareholders. The nuclear electric companies suddenly found they were competing in an increasingly price-sensitive market and, if they in turn were to successfully go the privatisation route, capital-intensive investment in new nuclear power stations was no longer an option.

The political climate also militated against any new nuclear build. The timescale from planning to commissioning a new plant was at least ten years. Thereafter, it might take twenty-five years to recoup the investment and return a reasonable profit. A five-year political cycle, with a Labour Party committed to policy of no new build, might stop a project in its tracks if Labour was returned to power. A newly built plant risked being mothballed, its expensive components sold off for scrap, as had happened with the Austrian Zwentendorf plant.

Privatisation of the electricity market also had implications for BNFL's flagship reprocessing project, Thorp. The original projections for Thorp – that half its throughput would come from UK power plants and half from overseas spent fuel – were no longer sustainable if no further nuclear power stations were built in Britain in the medium term. Further, its contractual arrangements for reprocessing AGR station fuel or the Sizewell PWR spent fuel output were far from secure. As the 1980s drew to a close, construction of Thorp was nearing completion. All going well, there should be no great difficulty in securing an operating licence for the plant once initial commissioning had been completed. Little did BNFL conceive that the most ferocious battle to save its flagship project from the nuclear scrap heap was just beginning.

18

Thorp

'Let us be clear about a number of things. THORP is located in another jurisdiction and is not answerable to the wishes of the Government or this House.'

Brian Cowen, Minister for Energy, 1994

BNFL set an initial target date of 1 January 1993 for the start-up of Thorp. The regulatory process meant there were several points along the path from construction to operation at which the project might be stalled or even aborted, points anti-Thorp campaigners were poised to exploit at every opportunity.

As well as having to meet European Community requirements under Article 37 of Euratom, that a nuclear plant in normal operation or accident conditions must not pose a threat to neighbouring states, licensing authorities in the UK could recommend that ministers 'call in' any BNFL application before granting the final go-ahead. Several rounds of public consultation might ensue before operating licences were granted. Economic justification provided grounds for a further public consultation, especially if the business environment had changed since the project was first mooted. Most feared by the industry, because of the prospect of delays of months or even years, was a whimsical minister disposed towards a public inquiry.

Unexpectedly, a potentially lethal threat to Thorp, and the nuclear industry generally, came from offside – an epidemiological study published in the *British Medical Journal* in February 1990. As part of the COMARE follow-up studies to the Black Report, Professor Martin Gardner and his team at the University of Southampton had examined fifty-two cases of leukaemia, twenty-two cases of non-Hodgkin's lymphoma and twenty-three of Hodgkin's disease that occurred among people aged under twenty-five years born in the West Cumbria health district between 1950 and 1985.

Gardner found a statistical association between the risk of childhood leukaemia and non-Hodgkin's lymphoma and the level of exposure to external sources of radiation experienced by fathers at Sellafield before the children were conceived. He claimed that this association could account for the Seascale cluster of these diseases. Gardner's hypothesis implied that even the strictly controlled and monitored dose limits allowed in nuclear plants heightened to a detectable level the risk of leukaemia in the children of fathers exposed to radiation.[1] By inference, the children of any male exposed to continuous doses of low-level radiation were similarly at risk.

Many epidemiologists, including Richard Doll, were immediately sceptical of Gardner's hypothesis. It ran counter to the long-range studies of the children of male atomic bomb survivors at Hiroshima and Nagasaki. Bomb survivors had on average received four times the lifetime radiation dose of the average nuclear worker and there was no evidence of excess leukaemia cases among their children. The counter-argument was that the atomic bomb survivors' dose was instantaneous, unlike cumulative low doses of radiation absorbed by nuclear workers over a prolonged period. Yet Gardner's hypothesis posed a further conundrum: the pre-paternal irradiation syndrome appeared to apply only to children born in Seascale. In the wider West Cumbrian community, which covered 93 per cent of the male working population at Sellafield with children, no such effect was discernible.[2]

Once again Seascale village played host to the media, including tabloid reporters and photographers who seemed particularly fixated on taking pictures of local children.

* * *

The Gardner report precipitated an immediate crisis at every level within BNFL. Jim Innes recalled a meeting with the shop stewards later that month prior to a presentation by Dr Gardner of his findings to the Sellafield workforce.

'They told me they didn't believe it,' Innes said. 'They didn't believe Gardner. Their own fathers had worked in the old Windscale. They knew the doses those men had received were far, far higher than anything they themselves might ever remotely be exposed to in the course of their work.'

Dr Roger Berry, who replaced Peter Mummery as BNFL Director of Health and Safety on the latter's retirement in 1987, unwittingly turned a crisis into near disaster. At a press conference following Gardner's session with the workforce on 21 February 1990 Berry was persistently asked by the *Whitehaven News* what advice he would offer site workers contemplating starting a family. It was a loaded question for which the BNFL Director was unprepared.

'It may be that the advice is, if you are so worried ... that you do not have a family,' Berry responded.[3]

'Nobody took a blind bit of notice at the time because it had followed all these other serious questions,' a BNFL executive who was present said later. At the end of the conference the *Whitehaven* reporter was overheard discussing what Berry actually meant with his national media colleagues. Berry, it was generally agreed, had told the workers: 'Don't have babies.'

Back at Risley, Harold Bolter compounded BNFL's difficulties. Unable to contact Berry directly, Bolter released a statement. Berry's comment was not company policy, he said. National trade union leaders, apoplectic at what they perceived as BNFL's cavalier attitude towards the Sellafield workforce, threatened industrial action. The unions looked for a meeting with the company chief executive, Neville Chamberlain. Forty-eight hours later, on 23 February 1990, the two sides reached agreement on lower radiation thresholds, well below existing national and international limits, for the Sellafield workforce. BNFL further offered to fund research into the Gardner hypothesis. Inadvertently that too would lead to controversy when Bolter, without prior permission, published the findings of a review by epidemiologist Dr Peter Smith that questioned the interpretation of Gardner's findings. Smith issued his own release stating he had provided the material to BNFL in confidence, generating further adverse publicity for the beleaguered company.

Within BNFL, there were many who thought nobody at senior level quite knew what to do. It was a situation that demanded cool heads, but there were few in evidence.

'They felt they'd invested loads of money in Thorp, they thought it was all going to be closed down. They felt the whole thing was just running away from them, that the company was arriving at its end with legal action being threatened,' one source said.

The personal injuries solicitors firm, Leigh Day and Co., had for some time been preparing files on West Cumbrian and Lancashire cancer cases for a class action suit of forty cases against BNFL. Solicitor Martyn Day informed BNFL that two cases would be brought forward. The first involved a leukaemia case, Dorothy Reay, who had died of the disease at the age of ten months in 1962, daughter of Sellafield worker, George and Mrs Elizabeth Reay of Whitehaven. The second, a lymphoma victim, Miss Vivienne Hope, whose father had worked at Sellafield, had survived her debilitating illness and was employed by BNFL.

The Hope & Reay case started in the High Court on 26 October 1992. The case, which continued for ninety sitting days until 26 June 1993, was extraordinary in that the hearings were dominated by expert scientific evidence on the reliability of Gardner's hypothesis. The ultimate judgement of the court was based on this evidence rather than points of law. Gardner's theory failed to convince Mr Justice French. Delivering his verdict on 8 October 1993, the judge stated: 'On the evidence before me, the scales tilt decisively in favour of the defendants [BNFL].'

Subsequent large-scale epidemiological studies, in Canada and the United States, as well as studies commissioned by COMARE in Britain, would provide no further support for Gardner's paternal preconception irradiation hypothesis as a factor in subsequent leukaemia among children.[4]

Anti-nuclear activists, however, persisted in their belief that the judge was mistaken in his interpretation of the evidence. 'The number of childhood leukaemia cases around Sellafield remains ten times higher than the national average,' Greenpeace said. 'BNFL has yet to prove that these facts are not connected.'[5]

* * *

Against this backdrop of damaging litigation proceedings, BNFL also risked failing to get a grip on the range of other potential derailments of the Thorp project. BNFL applied to the licensing authorities in April 1992 for new site discharge authorisations, for Thorp and the new enhanced actinide removal plant (EARP).[6] Sellafield management sat down with the licence authorities that July to review the timetable for the authorisations. The regulators told them an eight-week public consultation on the discharge authorisations would commence on 17 August that year. Already, Whitehall had received more than 12,000 letters, mainly orchestrated by Greenpeace. Government officials were informally warning the company that another public enquiry might become inevitable.

In early June, BNFL Corporate was itself unnerved to learn from the national press that the Sellafield management had granted permission to the local environment group Cumbrians Opposed to a Radioactive Environment (CORE) to stage an anti-Thorp demonstration at the site. Greenpeace had taken over the planned protest, now billed to include a rock concert by the internationally acclaimed

Irish band, U2, outside Sellafield's main gates. At least 15,000 young people were expected to attend from all over the north of England. Culture-starved Sellafield apprentices had even applied for the day off work. BNFL obtained an injunction prohibiting the use of its land for any such demonstration.

U2 was the star attraction at a Stop Sellafield gig in Manchester on Friday evening, 19 June. Afterwards, the band members, together with Greenpeace activists, secretly boarded the Greenpeace ship, *Solo*, and set out for the West Cumbrian coast. Reporters were ferried across the mountains to Seascale in transport supplied by Greenpeace.

At around 5 a.m. on Saturday morning the *Solo* appeared on the horizon. Bono and his band, together with about sixty Greenpeace campaigners, advanced to the shore in inflatables. Giant posters proclaimed: 'Nuclear power – a dead end'; 'React – stop Sellafield'.

Bono and band guitarist, The Edge, rolled barrels of mud ashore, taken from beaches along the Irish coast that they claimed were contaminated by Sellafield discharges. Their three-hour protest was staged on the high water mark, technically not on BNFL land and thus evading the terms of the injunction.

'It's absurd that a rock band has to dress up in ridiculous costumes to draw attention to this problem,' Bono was quoted.[7] The singer condemned the Irish government for its 'inaction' on the Thorp issue. Thorp would increase radioactive discharges from Sellafield a thousand-fold, Bono claimed. He would not allow his wife and two children to bathe in the Irish Sea.[8]

* * *

Greenpeace and Friends of the Earth supplemented public protests and letter campaigns with intensive lobbying at Whitehall and parliamentary committees at Westminster, questioning the economics of Thorp as well as its environmental impact. Accounts filtered back to the company that junior minister, David MacLean, had been persuaded by Greenpeace that Thorp would be a 'white elephant'.

It had never occurred to the BNFL commissioning team that politicians might turn against the Thorp project.

'Thorp was the only time that management called in the NCNI and said "We need your help",' Jim Innes recalled. The Sellafield shop stewards had been busy working the trade union vote and the Labour Party MPs. By 1992, Labour policy had softened from a commitment to phase out the industry to a promise of no new nuclear build.

The NCNI selected 'Trust Us' as the theme for their Thorp campaign and persuaded their own unions to subsidise it with four-figure sums. When they unveiled the strategy to the company's marketing and public affairs managers, a BNFL press officer said: 'That will never work.'

'And we asked: why not?' Innes recounted. 'And he said because they don't trust us, the British people don't trust us. And we said: No, they don't trust *you*, but they do trust us. They might not agree with us but they do trust us.'

As a counterbalance to Greenpeace, the NCNI 'Trust Us' campaign collected 10,000 signatures to a petition in favour of Thorp. The shop stewards embarked on a series of press conferences throughout Britain, starting in Norwich, constituency of Environment Secretary, John Gummer, who was known to be less than sanguine about Thorp.[9]

Wherever a regional media centre existed, the NCNI appeared. Cardiff and Bristol were followed by Birmingham, Manchester, Newcastle, Glasgow and Belfast. Channel 4 broadcast a documentary on their campaign. Dublin, virgin political territory to the shop stewards, was circled on the map as the final destination of their media tour.

* * *

The licensing authorities had reported to ministers in May 1993 that the draft authorisations, in their view, 'would protect human health, the safety of the food chain and the environment generally'. But there were wider policy issues, the inspectors said, that did not fall within the scope of their public consultation, such as the economic justification for Thorp.[10]

BNFL had put in place its own dedicated team of public affairs executives and consultants to take the fight for Thorp to Whitehall and Westminster. Based on an order book approaching £8bn, BNFL estimated Thorp would generate profits of £500m in its first ten years. Greenpeace said Thorp would produce a £600m loss.

BNFL seized on a civil servant's suggestion that an independent assessment of the economics of Thorp might help advance their political case. The accountancy firm Touche Ross was commissioned to conduct an independent assessment of Thorp's viability. Touche Ross reported Thorp would turn a profit of £550m. If the project was abandoned, £1.2bn in foreign earnings would be lost to Britain.[11] John Gummer told the Commons on 28 June 1993 that a further round of public consultation would take place, beginning on 4 August and concluding in early October. The consultation was based on documents supplied by BNFL on the economic and environmental case for Thorp and the proposed discharge authorisations.

This exercise generated 42,500 individual responses, including a second comprehensive submission from the Irish government that called for a public inquiry. Eighty-five local councils in Britain, and the Isle of Man, also sought a public inquiry. Overall, the responses were 63 per cent against the opening of Thorp, with 29 per cent, or about 12,300, demanding a public inquiry.[12]

Greenpeace Ireland had set out content guidelines for letters objecting to Thorp, purposely asking their supporters to write their own letters and sign them, to maximise both their authenticity and the unnerving effect on Whitehall. More than 10,000 letters from Ireland were wrapped up in a DHL parcel and sent to London.

The full Touche Ross report remained unpublished – to preserve commercial confidentiality, the UK government suggested. BNFL was strongly opposed to publication, fearing it must inevitably lead to yet another round of public consultation, and further delays to the start-up of Thorp. Friends of the Earth, probably in frustration at not being able to question the report's conclusions, would later wrongly claim that the Touche Ross report never actually existed.[13]

BNFL set April 1994 as Thorp's new commissioning date.

* * *

Dragged out through most of 1993, the Thorp campaign was punctuated by events that to those immediately involved appeared bizarre or sinister or sometimes both. In the late 1980s, Tom McLaughlan, BNFL's liaison officer with parliament, had been unsure how to interpret a letter, personally addressed to him and signed by Gerry Adams, stating that Sinn Féin in Northern Ireland was totally opposed to the construction of Thorp. Understandably, paranoia about the security of British nuclear sites from the IRA remained as strong as it had been in the 1950s.[14] McLaughlan alerted company security.

Now, the BNFL team was even more bemused by an unsolicited public statement from the Vatican press office declaring Pope John Paul II's principled support for nuclear reprocessing and recycling nuclear materials. At the height of the campaign, posters bearing the logo of the *Sun* newspaper popped up across London, emblazoned with the legend: 'The *Sun* says NO to Thorp'. BNFL public affairs rang the paper's editorial office. 'Hang on a minute,' said the voice on the other end of the line. The voice was then heard shouting across the newsroom: 'There's someone on here asking about Thorp. Whose side are we on?' Whatever side they were on, it later transpired that the posters were a hoax and had not been authorised by the newspaper. BNFL suspected anti-Thorp campaigners were to blame.

In December 1993, the BNFL team received a Christmas card from Greenpeace UK. McLaughlan returned the compliment. 'And thanks for all the overtime,' he wrote back.

Early in January that year, McLaughlan's colleague, Gavin Carter, had taken a phone call from Liam Lawlor, the Fianna Fáil deputy for Dublin West. BNFL should come to Dublin, Lawlor said.

Lawlor was one of a number of Fianna Fáil backbenchers who had long pursued the Sellafield issue, and maintained a close working relationship with John Bowler in Greenpeace. Now he offered to arrange a series of meeting with the Minister for Energy – his party colleague from Offaly, Brian Cowen – and representatives of all parties in the Oireachtas.

Led by Director David Bonser, BNFL arrived in Dublin on the morning of 23 February 1993. Bonser's message was straightforward: Thorp was a matter for decision by the British government. Greenpeace claims that Thorp would increase radioactive discharges from Sellafield a thousand fold were complete nonsense. While the site discharge of some radionuclides, such as tritium, and atmospheric emissions of radioactive krypton would increase significantly, the overall downward trend in discharges from Sellafield would continue, especially as BNFL's remediation plants, like EARP, came on line.

BNFL felt it had a strong financial case for the project. Backed up by the Article 37 opinion of the European Commission,[15] they felt they could realistically assert Thorp would have no detrimental effect on the Irish environment.

At their meeting with Brian Cowen, the minister left BNFL under no illusions about the strength of Irish feeling against Thorp. While there was no prospect of a meeting of minds between the visitors and the parliamentarians, Gavin Carter felt the dialogue was worthwhile and provided a basis for more constructive engagement with Ireland about nuclear issues in the future.

Unfortunately, on the day, BNFL's press office had separately agreed the delegation would attend a press conference with Lawlor in Buswell's hotel across the road from Leinster House. Bonser and his colleagues had to beat their way through a scrum of Greenpeace anti-nuclear pickets to gain access to the hotel. Inside, the atmosphere at the press conference was hardly any more cordial. As the delegation sped towards Dublin airport that evening in the heavily reinforced car of the British ambassador,[16] they felt, as one of them later described it, as though they were being run out of town in a tank.

Shortly afterwards the NCNI came to Dublin for their press event. 'Our attitude to the whole thing was that in Ireland they wouldn't know how many jobs are dependent on this and we'll go and tell them how important it is to our community and while we're at it, we'll find a way of letting them know how Irish our community is,' Jim Innes recalled. The shop stewards held a constructive meeting with the Irish Congress of Trade Unions who then accompanied them to a well-attended press conference.

The NCNI had not previously paid any attention to Ireland. Europe was where the grand nuclear battle was staged, where they lobbied MEPs and linked up with the nuclear workforce in Germany, France and Belgium to form TENS, an association of European nuclear workers, producing newsletters and pamphlets to make their case in the corridors and meeting rooms of the European Parliament in Brussels and Strasbourg. The shop stewards were surprised at the depth of hostility towards Sellafield when they first reviewed Irish media press clippings prior to their visit.

'A bunch of missionaries [Greenpeace] had walked in and converted the whole country,' Innes said. 'At least that's how it appeared to us.'

Following what they rated a good day's work, the stewards repaired to a public house in the centre city, and engaged in conversation with some young men at an adjoining table in the bar. When asked what their business was in Dublin they gave an honest reply. A few minutes later they were told they had better leave – for their own safety. Shocked and scared, they silently retreated to their hotel.

'We thought the worst thing we'd find in Ireland was that Greenpeace had been given a five year start,' Innes said. 'We had not the slightest idea that Sellafield might fit into another agenda of British imperialism or Irish neutrality or Republican freedom fighting. We simply had no idea.'

* * *

The first coalition government between Fianna Fáil and the Labour Party in the history of the Irish state had taken office in late 1992. A November general election delivered an inconclusive result that cast the Labour leader Dick Spring, with his thirty-three 'spring tide' Dáil seats in the role of king-maker. Spring opted for government with Albert Reynolds, who had succeeded Charles Haughey as Fianna Fáil leader in 1990 and spent two uncomfortable years in coalition with the PDs before finally falling out with its leader, Des O'Malley.

In the difficult choice facing the Labour leader between John Bruton's Fine Gael and Reynolds, Fianna Fáil offered the better deal on implementation of Labour policies. Reynolds further indicated he had secured a commitment at EU level to a massive tranche of European aid to repair and revitalise Ireland's woeful Victorian-standard infrastructure. In addition, he personally held out the tantalising prospect of a major development to end the Northern Ireland Troubles. In Reynolds' mind, resolution of the Northern Ireland Troubles and a take-off of Ireland's economy were inextricably linked.

The Downing Street Declaration jointly announced by the Taoiseach, Albert Reynolds, and Prime Minister, John Major, on 13 December 1993 reflected the efforts made by Reynolds, and others within the government, and groups on all sides of the conflict in the North, to map a way forward out of the political morass that had defined Ireland's history for decades.

Paragraph 4 stated: 'It is for the people of the island of Ireland alone, by agreement between the two parts respectively, to exercise their right of self-determination on the basis of consent, freely and concurrently given, North and South, to bring about a united Ireland, if that is their wish.' In time, this was enough for the Provisional IRA and loyalist paramilitary groups in Northern Ireland to declare ceasefires and lay down their arms and embark on a long and often fractured path towards participation in normal constitutional politics in the province.

Immediately following the Prime Minister's statement on the hopes for a negotiated peace in Northern Ireland, the Secretary of State for Environment, John Selwyn Gummer, rose in the House to announce that he and the Minister for Agriculture, Food and Fisheries had decided to grant the authorisations for Thorp and EARP. Subject to some minor conditions, the authorisations would take effect in twenty-eight days. Nothing likely to emerge at a public inquiry would significantly alter the basis on which the ministers' decision had been made, he said. When asked if he anticipated that the decision might cause dismay in the Republic of Ireland, Gummer simply noted: 'I would not have given the authorisations were I not sure that they properly protected public health and the environment.'[17]

In Dublin, there was a less benign interpretation of the back to back announcements of the Downing Street Declaration and the go-ahead for Thorp. The British government was accused of seeking to bury bad news on the day of an historic breakthrough on Northern Ireland. Gummer had displayed 'cynicism and grotesque *realpolitik*', one deputy charged. The opposition and government backbenchers angrily demanded immediate legal action against the British government. East coast local authorities, including Dublin County Council, which had engaged the services of sometime engineering consultant to Greenpeace, John Large, to prepare consultancy reports, 'would be willing to go ahead if the Minister were prepared to indemnify us,' a Labour backbencher offered.[18]

The timing of the Thorp announcement on the back of the Downing Street Declaration may have been more an unfortunate coincidence than a deliberate British snub to Irish nuclear sensitivities, although John Gummer, as Secretary of State, had a particularly bizarre sense of timing and diplomacy, as events in 1997 would later confirm. At the higher intergovernmental level there was no cross-over between the issues, according to Dr Martin Mansergh, who was a central advisory figure in the Northern talks process. As for the two announcements occurring on the same day, Mansergh said he wasn't even aware of it at the time.

* * *

On his third attempt, Trevor Sargent had been elected in 1992 as a Green Party deputy for Dublin North, a constituency he shared with Ray Burke, the national politician still most associated in the public mind with the anti-Sellafield campaign. Earnestly seeking to make his own mark, Sargent had once contrived to get himself arrested during a protest against plutonium transports and thereafter represented his brief detention by the West Cumbrian constabulary as the badge of honour among his anti-nuclear credentials. Like many of his parliamentary colleagues, lack of evidence or the scientific incoherence of any argument was no deterrent to its deployment.

Following the authorisation of Thorp, Sargent determined on a no-holds barred approach. The black-headed gulls of Ravenglass near Drigg at Sellafield were abandoning their nests because of radiation, Sargent told his Dáil colleagues during a debate on a motion he tabled in April 1994 demanding legal action against Sellafield.[19] Presumably the Dublin North deputy was unaware of a scientific study published three years earlier that showed the gulls were the least irradiated bird species around Sellafield. Too many foxes, too few rabbits and the dry summer of 1984 were the real culprits in the colony's temporary decline in the mid-1980s.[20]

The three-fold increase in emissions of krypton 85 from Thorp would have a deleterious effect on climate change, Sargent further claimed. He subscribed to a so-called 'superlinear' radiation theory, suggesting even the lowest doses of radiation would have fatal results.

Two American statisticians had postulated, among other things, that the origin of AIDS in Africa was connected to 1950s nuclear weapons tests fallout, due to the concentration of strontium 90 in human bones. For reasons of diet, they surmised, African bones were more susceptible than Asian or European bones to strontium 90. Ultimately, this manifested itself in the AIDS pandemic.[21]

'Perhaps I could draw the Minister's attention to a book published in the United States called *Deadly Deceit, Low Level Radiation, High Level Cover-Up*, in which two United States statisticians conclude that radiation is partially responsible for the spread of AIDS and has caused millions of excess deaths in the United States?' Sargent offered.[22]

Sargent was hardly unique among Irish parliamentarians in being either unwilling, or unable, to distinguish between lunatic conspiracy theories and real science. The difficulty for the Irish government was that in any court action real science would prevail. Attorneys General consistently advised there was no evidence of injury, environmental or health, to the Irish people from Sellafield, without which any case was unsustainable. Nor was there any evidence that Britain was in breach of any European law in the operation of its reprocessing business at Sellafield, including Thorp. The European Directive that required environmental impact assessments had only come into effect long after planning permission had already been granted for Thorp.

Greenpeace Ireland presented the department and ministers with their own blueprint for an international court case several times over from 1989 onwards. Ireland could seek redress against Britain through arbitration at the Paris Commission, Greenpeace urged, or a tribunal case under the United Nations Law of the Sea, or in the International Court of Justice or the European courts. Or if all else failed, in the UK courts.

The legal advice from the Attorney General's office was monotonously negative: any case would fail for lack of evidence. Politically, successive Ministers for Energy and up and coming political figures, such as Bertie Ahern and the Progressive Democrats' Mary Harney, also held to the line that a failed action might prove more damaging to Ireland's interests than not to take a case at all.

As for action specifically directed against Thorp, 'it is not expected that the consequences of the normal operation of the THORP plant would be such as to provide a basis for any legal action,' Brian Cowen said. A failed court action by Greenpeace and Lancashire County Council seeking a judicial review referred to the process of the ministerial decision to grant the Thorp authorisations, not operation of the plant itself, he pointed out.[23]

Cowen was anxious to quash any association between the government's refusal to initiate a court action against Britain on Thorp with the need to advance a peaceful settlement in Northern Ireland. 'It is wrong or mischievous to make non-existing connections or suggest that progress can be achieved in one at the expense of the other,' he said.[24] For the present, the Irish government would stick with its policy of direct diplomatic engagement with Britain on Sellafield and simultaneously exert whatever pressure it could internationally.

The 1992 Article 37 European Commission Opinion on Thorp had recommended the two countries conclude a more formal agreement on nuclear issues. Cowen had received a draft bilateral treaty from the UK, but judged its proposals 'inadequate'. He would hold out for a better, more 'wide-ranging' offer, he said.[25]

There were others to whom the government's caution appeared the real inadequacy: in March 1994, four residents of Dundalk, Co. Louth, brought a case to Dublin's High Court seeking an injunction against Thorp on the grounds that its operation was in breach of European law.[26] They also sued for damages for personal injury they claimed to have sustained from Sellafield pollution.

19

A Political Bone of Contention

> 'The people of Ireland have a legitimate interest in any proposal for a repository for radioactive waste near the Irish Sea coast.'
> John Selwyn Gummer, 1997

St Patrick's Day, 17 March 1997, and John Gummer's office was desperately searching for the Irish Minister of State for Energy, Emmet Stagg. The duty officer at the British embassy was instructed to track Stagg down on Ireland's national holiday, when officially the country was closed for business.

Gummer had news for Stagg, news he had determined to convey to him ahead of his own cabinet colleagues and as his Prime Minister, John Major, was on his way to Buckingham Palace to request the dissolution of parliament; ahead too, of the nuclear industry or any public statement. The Environment Secretary had decided to reject the Nirex application to construct a rock characterisation facility (RCF) at Gosforth near Sellafield as the preparatory stage to building a repository for Britain's nuclear waste.

The icing on the cake came in Gummer's official statement confirming the decision: 'The people of Ireland have a legitimate interest in any proposal for a repository for radioactive waste near the Irish Sea coast.' Britain, for the first time, Stagg claimed, had acknowledged Ireland's right to consultation on nuclear issues.[1]

It was a rare political moment to be savoured by the Rainbow administration of Labour, Fine Gael and the Democratic Left parties that installed John Bruton as Taoiseach without a general election following the collapse in November 1994 of Labour's brief power-sharing experiment with Fianna Fáil.

A breakdown in the Provisional IRA ceasefire in February 1996 had left the northern peace process stranded. John Major's dependence on the unionist vote in Westminster to prop up his government compounded John Bruton's difficulties in striving to find a way out of the impasse, and the new Taoiseach was perceived as floundering on Northern Ireland. The Irish economy showed all the signs of a dramatic growth phase, marked by a sharp fall in unemployment, hi-tech industrial investment and the start of a property boom, in the face of which the government's economic strategy appeared cautious and lacking in creativity. Even on social policy, while a referendum to allow civil divorce was passed if only by the narrowest of margins in 1996, some ministers seemed hidebound by misplaced loyalty to entrenched vested interests in their initial response to such issues as the exposure of clerical child abuse in state institutions. Historically, Irish political culture always relied more on impressions than substance. The Rainbow might lay claim to a record of competence in government but they faced a sullen electorate, whose voting intentions in the forthcoming general election remained inscrutable.

* * *

Stagg's victory over Britain's nuclear establishment was presented as a political triumph for a resolute man of action. In truth the failure of Nirex was largely attributable to their own incompetence in making the case for the RCF with their peers, the UK planning authorities, the British public and domestic political factors. Refusing the RCF reflected both John Gummer's personal antipathy towards the British nuclear industry and the weakness of a Conservative government facing a general election they were bound to lose.

For decades, British governments had allowed grain-style silos and sludge tanks on the Sellafield site to fill up as an alternative to developing any strategic long-term solution to nuclear waste. A national repository was a national necessity: as far back as 1976 the Royal Commission Flowers Report recommended no further nuclear stations should be built in Britain until the waste issue was resolved.[2] But in the eyes of successive Secretaries of State, files on a nuclear waste repository had otherwise invisible 'not in my political lifetime' stamps all over them.

A national repository for low-level nuclear wastes already existed at Drigg. Safeguarding high-level reprocessing waste, much of it due for return to overseas customers, was BNFL's direct responsibility. From 1982, Nirex, in which BNFL had a one-third stake and owned jointly with the nuclear generating companies, had the sole task of finding a suitable long-term underground storage base for intermediate level nuclear waste (ILW).

'Long term' meant anything from a few hundred years to thousands of years before the radioactivity in the stored material decayed to normal background levels in the environment. Within five years Nirex had identified four possible sites, generating community uproar in each area they were announced. Two weeks prior to the 1987 general election, the sites were withdrawn.

By comparison with Britain, Ireland's radioactive waste problem was minor, but the concept of a permanent storage base raised similar political issues in both countries. Large or small, on the surface or deep underground, a repository would just sit wherever it was put, probably for millennia. Unlike the electricity from power stations or the employment and financial returns from commercial reprocessing, a repository produced nothing of value to its host community. Theoretically, properly planned and executed, it should do no damage either.

Nirex resisted pressure from environmental NGOs and the media to disclose their full list of prospective sites, but Dounreay or Sellafield were obvious contenders.[3] West Cumbria appeared the logical choice since two-thirds of Britain's ILW was already concentrated on the Sellafield site. Nirex applied to Cumbria County Council in 1994 seeking planning permission for its laboratory at Gosforth, 1,000 feet underground below the water table, to test geological formations in the area. The council refused to believe that the RCF, construction of which would cost up to £200m, was not a precursor to a final repository in West Cumbria, and turned down the application. Nirex appealed.

The appeal, heard at Cleator Moor over a four-month period from January 1996, was the first time an Irish government minister attended a British planning inquiry.[4] Emmet Stagg personally led a delegation of Irish officials to the hearing on 12 January 1996. The Irish government's objection, based on fears that nuclear materials might leach from the repository to the Irish Sea and cause economic damage to Ireland's fishing industry, was outlined by Elihu Lauterpacht QC, a distinguished and internationally acclaimed specialist in public international law, and advisor to the Irish government on its international case strategy, and James Hamilton, head of the Attorney General's office. The sharp intakes of breath in Whitehall, and throughout the nuclear industry, when it was first learned that Stagg proposed to attend the local planning inquiry in West Cumbria's 'Little Ireland', turned to sighs of relief when the intervention passed off with dignity and decorum.

Nirex also failed to convince the planning judge of their scientific rationale for the RCF. Media speculation, however, suggested a final decision, which rested with the Secretary of State, might be postponed until after the next general election. There were further sharp intakes of breath when a planned meeting with the British ambassador to Ireland, Veronica Sutherland, on 7 February 1997, was represented in parts of the Dublin media as the Irish minister having 'called in' the ambassador, a phrase in normal diplomatic parlance redolent of a formal reprimand. Early in March Stagg made a further submission to the British authorities detailing Ireland's objections. If permission for the RCF were granted, it would create a potential source of conflict and the Irish government would consider initiating legal action to stop it, he told the media. 'This will be a bone of contention between these islands which share a common boundary, the Irish Sea,' Stagg said.[5]

Stagg did not lack company claiming credit for the knock-out punch to Britain's nuclear industry. Irish MEPs and representatives of all parties and anti-nuclear groups queued to stake a claim in this 'historic victory'.[6]

To many in BNFL, who had long despaired of Nirex's approach – the secrecy with which it guarded the site list, and its apparently dismissive response to questions regarding the geological suitability of the Gosforth site for its RCF – the Nirex failure was predictable, if unfortunate for the industry as a whole. Nirex had alienated public opinion in West Cumbria, including the Sellafield trades unions, who expressed their own concerns about the quality of Nirex scientific data and long-term safety of a repository.[7]

BNFL repositioned the waste issue as a political rather than a technological problem. BNFL had always maintained a discreet arm's-length relationship with Nirex. Now they effectively abandoned them. BNFL stated there was ample overground storage space at Sellafield for ILW, for the next fifty years if required.[8] The crux was that any decision on the final waste disposal route should be the right one and be broadly publicly acceptable. The company could point to Sweden, Finland and France as countries that, albeit taking an approach that sought to reconcile the national requirement for a storage depot with the interests and views of local communities, were making progress towards solving the problem of long-term storage of nuclear waste.[9]

* * *

BNFL public affairs executives calculated that if a similar ratio of objections to Thorp as came from Ireland had occurred in Britain, then the second public consultation might have generated 250,000 objections instead of 42,500. Public affairs manager Rupert Wilcox Baker began to work on a strategy to challenge the 'fear agenda' he believed dominated Ireland's anti-nuclear policy. Aware that Britain's first draft for a bilateral treaty on nuclear issues was rejected by Brian Cowen at the height of the Thorp campaign, Wilcox Baker anticipated it could take several years before a political climate conducive to its successful conclusion emerged, given the political atmosphere post-Thorp and the inception of the Louth court case. Irish governments, he correctly predicted, would at best continue to stall on any proposal from Britain for a bilateral agreement on nuclear issues.[10]

BNFL could support the process by actively engaging with the media and in political and public debate at all levels in Ireland; challenging misinformation, issue by issue, from safety at Sellafield, to the environmental and health impact of discharges on Ireland, to the security and safety of nuclear transports through the Irish Sea. A handful of independent Irish scientists and academics, and the RPII, regularly sought to set nuclear issues in context, but dull science and complex data made for poor competition with high-flown political rhetoric. If only because of their status as 'public enemy No. 1' in Wilcox Baker's estimation, BNFL could hardly be ignored by the Irish media.

Unusually within the British nuclear establishment and BNFL itself, Wilcox Baker held to a strong personal belief that BNFL had a corporate responsibility to take its case to Britain's nearest neighbour, however unpleasant the experience might prove to those immediately engaged in it. BNFL also had a strong vested interest in Ireland in terms of managing its international reputation, particularly in the context of the company's aspirations to expansion and privatisation.

BNFL chairman, John Guinness, a former Secretary of the Department of Energy,[11] needed little persuasion. Through his connections to both the brewing and banking sides of the famous Dublin family, he was sympathetically disposed towards Ireland, even if sometimes personally perplexed by its anti-nuclear politics.

Apart from Ireland, Guinness, with his formidable though engaging personality, had a difficult agenda to see through. The 1971 Act establishing BNFL allowed for a sale of 49 per cent of the company's shares without recourse to parliament. A first stage sell-off, the creation of a public–private partnership (PPP) was anticipated at the latest by 2002. From the early 1990s pressure intensified within BNFL to prepare for a privatised future. BNFL's transition from a national centre for managing Britain's nuclear waste to a profitable, internationally trading, 'cradle to grave' nuclear services company was a prerequisite to successful privatisation, whether partial, a PPP, or a full sale.

John Taylor, who had spent the previous seventeen years of his career with Exxon, was appointed CEO of BNFL in 1996 to implement radical organisational change. In practice this meant expansion into new services, such as decommissioning, gaining a foothold in the reactor services and new-build markets and changing the way in which BNFL, and Sellafield in particular, went about their business.

The West had temporarily abandoned nuclear power, but new markets were opening up in Korea and China. As international concern over global warming increased, BNFL evaluated nuclear's future potential as a carbon-free electricity source. Within a decade, a resurgence of nuclear power was not just likely, but inevitable, they concluded. BNFL acquired Westinghouse nuclear services division in 1999, and with it, access to new reactor designs.

American expertise in radiation packaging technologies lagged far behind the British and the French, partly due to the presidential moratorium on reprocessing in the 1970s.[12] With the end of the Cold War, significant business opportunities for clean-up decommissioning, in which BNFL already had specialist expertise from Sellafield, emerged at old military sites in the US and to a more limited extent in Russia.

The Sellafield unions distrusted Guinness and Taylor. Guinness, they believed, was sent into BNFL to accelerate privatisation, which they instinctively feared would sound the death knell for reprocessing, their jobs and their community. Taylor was his hired gun. A partially privatised BNFL might be expected to promote its front end fuel and reactor services and decommissioning expertise in the international markets at the expense of reprocessing. New employment contracts that dramatically cut overtime, and the announcement of redundancies at Sellafield, confirmed their worst fears about where the company was heading. The unions were resolutely opposed to the PPP and sought to persuade their allies in the Labour Party to oppose it.

The unions were aware of the diminishing status within BNFL of reprocessing as a core business. For a start, the economics of reprocessing seemed less attractive in the mid-1990s than twenty years earlier when Thorp was first planned. Environmentally, Sellafield made no small contribution to Britain's image as 'the dirty man of Europe', and the site remained a magnet to protestors and anti-nuclear NGOs.

At government level, ministers and secretaries of state, including John Gummer at the height of the Thorp campaign, had to suffer the indignity of personal abuse from their international counterparts, as Irish concerns about Sellafield safety and discharges spread to Scandinavian countries like Norway, Denmark and Iceland.

Meanwhile, BNFL was reluctantly about to increase its own nuclear generating capacity beyond the small reactors at Calder Hall and Chapelcross. Too late, in the opinion of some industry experts, the nuclear electric companies, as British Energy, were privatised in 1996, just as Britain's 'dash for gas' and, with it, a rapid fall in wholesale electricity prices was gaining momentum. Vast improvements in productivity in the AGR power stations coupled with strong prices for electricity in the pool market meant that the nuclear generators had turned a handsome profit in the early 1990s. On market advice, the first generation magnox stations, with their uncertain liabilities and limited lifetimes, were excluded from the 1996 sale.

To the consternation of Guinness and the board of BNFL, the integration of BNFL and Magnox Electric was announced in December 1997 against the company's wishes and Magnox Electric became a subsidiary on 30 January 1998.

Taking over nine magnox stations, three of which were already closed and awaiting decommissioning, introduced further uncertainty into BNFL's liabilities projections and its privatisation plans. The merger also changed the dynamics of Sellafield fuel reprocessing contracts, since BNFL could hardly make profits from trading with itself.

* * *

Sellafield, as ever, was set to generate several new international controversies for BNFL and the British government. During the battle for Thorp the company held back from seeking planning permission to construct a new fuel plant, attached to Thorp, to manufacture the plutonium recovered from reprocessing into mixed oxide (MOX) fuel.

MOX completed the nuclear fuel services cycle. Under UK law, and international inter-government agreements, reprocessing contracts specified the return of its by-products, plutonium and uranium oxides and wastes, to Thorp overseas customers. Returning plutonium as MOX rather than raw powder was the preferred option, BNFL claimed, not just of BNFL's foreign customers but of international authorities, for obvious safety and security reasons. The planning application for the Sellafield MOX plant was presented to Cumbria's local authorities in 1994.

Scientists at Sellafield, preparing in 1996 for the next cycle of discharge authorisations for the site, drew BNFL Corporate's attention to an unanticipated problem with EARP. In the heat of the Thorp campaign, scant attention had been paid to the EARP discharge authorisation, which granted an increase from 20 terabecquerels of technetium 99 (Tc99) to 200 terabecquerels per annum. The 1996 application proposed reducing the limit to 150 terabecquerels.

The problem was that while the plant was highly successful in removing plutonium and americium from Irish Sea discharges, its filtering mechanism had no effect on Tc99, a weak beta radiation emitter, present in the medium active concentrate waste (MAC) from reprocessed magnox fuel. BNFL held back large quantities of MAC in storage pending the construction and authorisation of EARP. Now the MAC was being put through EARP in batches.

Technetium accumulated in crustaceans, especially lobsters, and in local samples was doing so at sixteen times the European Union threshold of 1,250 becquerels per kilogramme for radioactivity in food. The post-Chernobyl guidelines for radioactive contamination, CFILS, applicable to a basket of foodstuffs in the event of another large-scale accident, were drawn up and periodically reviewed by the EU Article 31 Committee. This expert group was charged with the task of imposing coherence on Europe's otherwise haphazard and ad-hoc national rules that had caused havoc and public confusion, as well as unfair bans on food imports from Hungary and Poland, in the aftermath of the Chernobyl accident.[13]

Given that Sellafield's effluent, drawn by natural currents in the Irish Sea, snaked around the Scottish coast and dispersed into the North Sea, it was only a matter of time before Tc 99 was detected in marine samples in Norway's fishing grounds. As to the likely severity of the contamination in lobsters landed in Ireland or Norway, the scientists were unsure.

The storage of highly active liquid waste in twenty-one tanks at Sellafield was also moving up the political agenda.

'The biggest threat to Ireland in terms of the UK nuclear installations is ... a catastrophic accident,' Stagg had told the Seanad in 1996. If the tanks' cooling system failed, then in 'just eight hours the resulting pollution would be equivalent to 100 Chernobyls 100 miles from Dublin'.

'If the breeze was blowing in the wrong direction from our point of view at that time, agriculture in Ireland would be wiped out forever; scientists say it would be for a period of 100,000 years.'[14]

BNFL privately invited the RPII to send representatives to Sellafield to visit the HAL facility and hold discussions with the plant operators. Following a brief visit in July 1997, Frank Turvey wrote to the company: 'We found the visit extremely useful. We were very impressed by what we saw.'

The RPII board, however, remained under sustained political pressure to demand a much more extensive verification of the safety of Sellafield's HAL storage facility, specifically its safety case. This was open to interpretation as the regulatory authority of one country seeking to pass judgment on the work of another, a proposition entirely unacceptable to the Nuclear Installations Inspectorate of the HSE or to the Department of Trade and Industry, irrespective of BNFL's inclination towards a more flexible response to RPII demands. BNFL won out, and an RPII team spent two weeks at Sellafield in early February 2000.

* * *

Rupert Wilcox Baker had no illusions about converting anyone in Ireland to nuclear power. If he had, an ill-judged BNFL Press Office project in 1995 would have entirely dissipated them. BNFL was approached by the marketing division of the *Sunday Tribune* newspaper and made a deal for production of a special supplement on Sellafield at a cost of about £35,000 sterling. Among its more flamboyant claims, the supplement used imperialist language to describe Thorp as 'the jewel in the crown' of Britain's nuclear industry. Publication almost caused a strike among the paper's journalists, who forced their editor to publish a front page disclaimer dissociating reporters from its production. The affair generated yet another negative round of news coverage on the general theme of BNFL's arrogance.

BNFL public affairs executives were familiar with Irish political sensitivities from discussions with members of the Dáil and Seanad during fact-finding visits to Sellafield, before and during the Thorp campaign, under the auspices of the British–Irish Inter-Parliamentary Body.[15] On one such visit in November 1993, Louth Fianna Fáil deputy, Dermot Ahern, advised his hosts that even if Sellafield's activities proved harmless to Ireland, politically they were unacceptable. On his return, Ahern spoke of being more frightened during his visit than he had been before he went there, such a useful tactical exposition for domestic political consumption that it was frequently replicated by other Irish parliamentary visitors to the site down through the years.[16]

At a meeting between Fianna Fáil and a BNFL delegation in 1996, party leader Bertie Ahern gave John Guinness a lesson in Ireland's electoral system: the vagaries of Ireland's single transferable vote meant no party facing into a general election could ignore the strength of popular feeling on any issue, irrespective of where it ranked on the list of public concerns. A fifth preference vote for a Fianna Fáil candidate ahead of a candidate for another party might ultimately mean winning or losing the last seat in Ireland's multi-seat constituencies, and hence the difference between being in government or on the opposition benches. If an anti-Sellafield policy would secure that fifth preference vote, then politics would dictate what the policy was.

At national level, the competition between politicians as they wrestled one another for the moral high ground on Sellafield was at times reminiscent of the pantomime children's game, 'Who's the king of the castle?' The Rainbow government had three ministers with a specific anti-Sellafield brief: as a former aide later admitted, the trio often rivalled one another as to who would be first to get out the latest media statement condemning Sellafield.

The Rainbow government broadened the Irish anti-nuclear agenda to one of undiscriminating opposition to all forms of nuclear power, not just reprocessing at Sellafield. Stagg was a member of CND; his counterpart in the Department of the Marine, Democratic Left's Eamon Gilmore, a member of Greenpeace. Gilmore introduced legislation to prevent ships carrying nuclear materials from docking in Irish ports without RPII consent. Initially he proposed banning any ships carrying nuclear materials from Irish ports, but this was prevented by the Department of Foreign Affairs, concerned about Ireland's international obligations, which forced an amendment to the Bill in its passage through the Seanad.[17] The senior member of this triumvirate, Labour's Brendan Howlin in Environment, established a ministerial task force in April 1995 to co-ordinate the efforts of all departments in the fight against Sellafield and liaise with the 'Louth Four' on the progress of their case.

For all its campaigning, including directions to Irish diplomatic missions to register Ireland's objections to Sellafield in countries[18] whose utility companies had Thorp contracts, the Rainbow coalition equivocated on its programme commitment to take an international legal case.[19] Fine Gael Junior Minister for Environment, and task force member, Avril Doyle, iterated the by now familiar line that Ireland 'cannot initiate such action without a firm legal case based on sound scientific evidence'.

'Worse than not taking a case would be to take a case prematurely,' she told a Sellafield conference in Drogheda in January 1996.[20]

* * *

The conference organiser, then Mayor of Drogheda, Fine Gael's Fergus O'Dowd, had contacted BNFL in late 1995 inviting the company to participate. O'Dowd believed he received a positive response from BNFL and released a statement to local newspapers. Shortly afterwards he received a letter from Rupert Wilcox Baker, politely declining the invitation.

When O'Dowd met Wilcox Baker in Dublin in late 1996 to discuss the possibility of BNFL's participation in a second Sellafield conference, his still seething anger was apparent as he drew the letter from his inside jacket pocket and placed it on the table between them.

'I still have that letter at home somewhere. Basically, what it said was that the issue was none of my business,' O'Dowd would recall over ten years later.

As BNFL, in some trepidation – O'Dowd's political rival, Labour Party deputy, Michael Bell, had threatened to picket the conference in protest at BNFL's presence – drove into Drogheda on the morning of 11 January 1997, the most they could hope was that their willingness to engage with a hostile public might lay foundations for further constructive exchanges.[21]

Coincidentally, less than a week later, Greenpeace International announced the closure of its Irish operation.

* * *

Following the establishment of Greenpeace International in 1979, Greenpeace was organised along the lines of a multinational franchise operation. National branches developed their own subscription base. They were subject to central directives and, at the very least, required to break even financially. Publicity was the oxygen that kept 'God's Navy' afloat and subscriptions rolling in, but by the mid-1990s the air supply was running thin and Greenpeace was in crisis.

Despite its decade of investment in an Irish national office, Greenpeace had failed in its objective, first declared in 1979, to close Sellafield. While Greenpeace Ireland had over 4,500 Irish members by

1996, it was running at a loss. The Irish office required a subsidy of £300,000 from its parent in 1996 to keep going, Greenpeace International claimed in the media.[22] John Bowler put the figure for 1996 closer to £50,000. Although he later said he understood the reasoning behind the closure decision within the context of Greepeace International's future strategy to develop the organisation, personally it was a sad moment. Greenpeace Ireland's budget projections for 1997, Bowler believed, would have delivered a break-even position.

A personal plea from Emmet Stagg – 'I wish to place on record my support for their retention in our country', he told the *Irish Times*[23] – proved to no avail. Unprofitable offices had to close, Greenpeace International insisted, and Ireland would be the first to go. The decision to shut down and disband Greenpeace Ireland was 'tough and unfortunate' but unavoidable.[24]

The 1995 anti-Shell campaign to force disposal of the Brent Spar oil rig on land instead of at sea delivered a temporary respite to Greenpeace's declining fortunes worldwide.

It also illustrated Greenpeace's occasionally shaky relationship with science. Science was important, but compelling evidence was entirely subsidiary to the ethical principle, as defined by Greenpeace, of any campaign. Scientific hyperbole, however outrageous or ill founded, exaggerating the theoretical possibilities of disastrous consequences of everything that Greenpeace opposed, from nuclear tests to GM food trials, was acceptable if it enhanced its ethical case.

The phenomenal PR success of the Brent Spar campaign was tempered by the use of false data on the toxicity of the contents of the redundant platform, a mistake for which Greenpeace had to publicly apologise. An advertisement in 1994, illustrated by a photograph of a baby with a grossly deformed head, claimed Thorp discharges would cause 2,000 deaths in ten years. Subsequent investigation by the British Advertising Standards Agency found the child's condition was not caused by radiation and the 2,000 deaths claim was unsubstantiated. Greenpeace International took out advertisements in four national newspapers in Holland, claiming a raised incidence of stillbirths around Sellafield. BNFL took action in the Dutch courts. Greenpeace was ordered to publish a retraction of their claims in the newspapers and were fined when they subsequently failed to do so.

The logic of Greenpeace's mixing of scientific claims and propaganda was impeccable: whenever its grasp of science was found wanting, Greenpeace could hastily revert to its ethical position:[25] sea dumping was inherently bad, nuclear power inherently dangerous.

The premier brand of the international environmental movement, Greenpeace's own administrative structure curiously aped many of the multinational targets of its most spectacular campaigns. By the very fact of eschewing broader political considerations in favour of a single-issue environmental focus, Greenpeace could dominate the national and international political agenda on any issue of its choice. Arguably, by 1997 the organisation no longer needed an office in Dublin; its job in Ireland was already well done.

* * *

The Louth campaigners too needed funds and scientific evidence to pursue the case they had initiated against BNFL, to which the Irish government and Attorney General were joined for technical legal reasons. They were billed as 'ordinary heroes' in the media, and their first major success in the case came with an apparent defeat of BNFL in the Supreme Court. While the parties were all agreed the personal damages claim could go ahead, BNFL had appealed an earlier High Court ruling that upheld the jurisdiction of the Irish courts to decide the European law aspects of the case.

The Supreme Court ruled on 24 October 1996 that the trial judge could decide on the jurisdiction issue when the case eventually came to trial. The sting in the tail of this judgment was that the case would remain locked within the Irish court system, rather than progressing to a grandstand hearing in the European Court of Justice (ECJ) at which BNFL's alleged mere lip-service to European environmental law could be exposed. It was well known in political and media circles that several leading Irish senior barristers were ready and willing to provide *pro bono* representation for the Louth Four on environmental law issues at the ECJ.

The case once again fixed the media spotlight on the unresolved mystery of the Dundalk Down's syndrome cluster. Dr Patricia Sheehan, who first identified the cluster in 1983, had died in a car accident six years later. Local Dundalk GP, Dr Mary Grehan, a close friend, was as convinced that the circumstances of the accident were suspicious as she was that the Windscale fire was implicated in the cluster.[26] Grehan, scheduled to appear as an expert witness in the Louth case, claimed to have evidence from Dr Sheehan's papers there were many other, as yet undocumented, Down's cases in the Dundalk area as well as a higher than average incidence of stillbirths. Surveys within her own practice showed vitamin B12 deficiency was rampant in Co. Louth, consistent, Dr Grehan feared, with exposure to low-level radiation. 'The whole place is radioactive, we probably glow in the dark by now,' Grehan said of her native Dundalk. 'This is fact – this entire area is contaminated and if someone such as BNFL wants to prove otherwise let them do the research.'[27]

The Louth case provided a campaigning platform to politicians of all parties. Nuala Ahern, a native of the county and a seasoned antinuclear campaigner long before her election in 1994 as a Green Party MEP for Leinster, pointed to an apparent excess of 14 per cent in Co. Louth over the national average incidence of cancer as evidence of the effects of low-level Sellafield radiation, when the National Cancer Registry (NCR) published its first report in 1996.

Arthur Morgan, later elected a Louth TD for Sinn Féin in 2002, had helped found a health group in Cooley. Residents of the picturesque north Louth peninsula were increasingly alarmed by the high number of young people in the area apparently falling victim to cancer. Morgan would brook no argument, even from the head of the NCR, Dr Harry Comber, that suggested that factors other than Sellafield radiation, such as local diet or smoking, might be responsible for a higher than average incidence of some cancers in Dundalk and the surrounding districts.

'We looked at the number of cancers in County Louth and the number that you might expect given the age and sex make up of the population,' Dr Comber pointed out in a local radio debate with the Sinn Féin councillor.[28]

Twenty-five more cancers in Co. Louth had been diagnosed in each of the years 1994 and 1995 than might have been expected in the population.

'If you look at the cancers that made up that excess of fifty cases, what you find is that the cancers are lung cancers, stomach, bladder, oesophagus cancer, kidney cancer,' Dr Comber added. 'None of these have ever been shown to be radiation related, other than lung cancer in the case of radon gas, but in terms of external radiation or food radiation or anything like that, there is no appreciable link between any of these cancers and external radiation.'[29]

Arthur Morgan accused the registry of covering up the facts. Morgan was joined by most other local politicians, including Fergus O'Dowd, in demanding further investigations into the locally high cancer rates and other health problems in Louth.

The North Eastern Health Board had already recruited Dr Geoffrey Dean, a former chief of the Health Research Board, to continue Dr Sheehan's research on the Dundalk Down's syndrome cluster.

* * *

The Louth case had potential to disrupt the Irish government's anti-Sellafield strategy. A well-intentioned motion by Labour Senator Pat Magnier on 30 October 1996 to applaud his ministerial colleague on the successes of his anti-Sellafield policy and celebrate the Louth Four's Supreme Court victory was spoiled when Fianna Fáil senators slipped in an addendum committing the Rainbow government to fully fund the Louth case, an undertaking already included in their own party's environment policy.

Stagg had no political alternative but to accept the Fianna Fáil amendment. The government had offered £100,000 to the Louth Four to fund research. Three days before the June 1997 general election, Brendan Howlin, in his capacity as head of the ministerial task force, upped the offer to £200,000. Publicly, Stagg indicated the government might provide more funds if the Louth Four dropped their suit against the state. Privately, the government was aware that the Attorney General's advice to Brian Cowen in 1994 was the same advice they were receiving: the case had no possibility of success. Any government had to walk a fine line on the level of support provided to the litigants. If a case whose legal costs were funded by the government was lost then BNFL might, in turn, embarrassingly lay claim to all its own costs from the Irish state.[30]

John Bruton's 1997 general election campaign tour of the country began with a visit to Drogheda.[31] The Taoiseach promised to back the Louth residents in their bid to close Sellafield. The resources of all state agencies would be thrown open to them and he would raise the Sellafield issue with Britain's new Prime Minister, Tony Blair, he said. For all the high-profile hysteria, when it came to casting votes, Sellafield remained well down the electorate's order of priorities. When the twenty-first Dáil convened for business on 7 July that year, Bertie Ahern was elected to replace Bruton as Taoiseach.

20

Friends and Neighbours

> 'Wicklow men have a reputation for being good at putting in the boot and I am no exception in that regard.'
>
> Joe Jacob on Sellafield, February 2000

The signing of the Good Friday Agreement on 10 April 1998 and its overwhelming endorsement in a referendum, North and South, just over a month later, forged a final step towards resolution of the real bone of contention between Ireland and Britain: a political settlement to the northern question.

The relationship between Britain and Ireland had advanced to a level of co-operation and understanding that would have left the founders of the Irish state incredulous.

'We're no longer obsessed by our relationship with Britain,' Fergus O'Dowd characterised this transformation in Anglo-Irish relations of the last quarter of the twentieth century.

Ireland and Britain could find common ground on most issues; the European Union, trade and commercial priorities and, most crucially, the way forward in Northern Ireland. The Irish and British governments probably worked more closely together across a range of issues than any other two governments in Europe. Sellafield was the exception.

Yet it had no impact on the overall relationship even when raised directly by Bertie Ahern with Tony Blair, Martin Mansergh insists.

'The relationship was between professional politicians and they understood each other,' Mansergh said. Agreement on Sellafield was about agreeing to disagree.

Former Energy Minister Brian Wilson was more forthright: 'The idea that Bertie Ahern and Tony Blair would have fallen out over Sellafield is just ridiculous.'

'There were public postures and private postures,' Wilson said, 'and I'm sure representations were always being made about Sellafield, but not in the same tone as the public things that were being said elsewhere.'

The significance assigned to Sellafield by those dealing with the issue was directly proportionate to their vested interests, whether they were the media, national politicians, NGOs or Her Majesty's diplomatic representation in Dublin. To the British embassy, Sellafield was an unwelcome irritation with potential to disrupt the otherwise smooth flow of Anglo-Irish relations, or 'east–west' as they were now termed. Ireland's economic boom in the late 1990s attracted unprecedented levels of British investment into Ireland, especially in the retail sector. The negative image of Britain's nuclear industry in Ireland and its predominance in Irish media coverage of Britain in Ireland was counterproductive to the embassy's promotion of commercial and cultural ties between the two countries.

The embassy alternately castigated BNFL for raising the profile of nuclear issues in Ireland by their insistence on engaging in public debate on Sellafield, or upbraided them for not doing enough to assuage Irish concerns. As far as BNFL public affairs executives were concerned, the embassy's position veered bewilderingly between metaphorical threats to send gunboats up the Liffey and complete appeasement of the anti-nuclear agenda of Irish politicians. BNFL was disconcerted by reports from its Irish political contacts of diplomatic whispers that in demanding the closure of Sellafield Irish politicians were 'pushing an open door' with the new Labour government.

Neither government nor opposition figures believed any diplomatic barriers existed to how they expressed concerns about Sellafield, whatever the issue: discharges, the risk of an accident or the transport of nuclear materials through the Irish Sea. Nor did any words or actions, however injudicious or extreme, carry any political consequences, either domestically, where they were more likely than not to be lauded in the Irish media, or within the broader context of Anglo-Irish relations.

At a time when Ireland was persistently threatened with prosecution by the European Commission for its stewardship of the environment, or rather the lack of it, Irish ministers, including the Taoiseach, Bertie Ahern, consistently identified Sellafield as the greatest environmental threat facing the Irish people. Based on a presentation by the anti-nuclear scientist, Dr Gordon Thompson, Emmet Stagg was encouraged to raise his estimate of the likely impact of an explosion at Sellafield's HAL facility from 100 to '200 Chernobyls'.[1] Irish politicians endorsed Thompson's 'grand bargain' proposal, under which all commercial activities at Sellafield would cease, with its workers redeployed to clean up nuclear sites at home and abroad.

When it came to energy policy, any lingering distinction between opposition to the environmental consequences of Sellafield reprocessing and a neutral stance on the use of nuclear for electricity generation had been long since abandoned by all parties. Under pressure from the Greens' Trevor Sargent and the Labour Party, the government included a provision in the Electricity Regulation Act, 1999 to ban nuclear power generation in Ireland.

'By opening up the European electricity market we may be supplied by French or British companies which draw their energy from nuclear sources,' Sargent had argued during the debate on the Bill. 'It is vitally important that there are strong rules which make it clear that we do not want to import electricity produced in such a hazardous manner.'[2]

A ban on the importation of nuclear-generated electricity to Ireland was in breach of EU rules on the free movement of goods and services. It was also impractical: electricity is simply electricity and it is impossible to discriminate as to its original source. The Republic was probably already importing nuclear-generated power from Scotland via the interconnector with Northern Ireland. All Minister for Public Enterprise, Mary O'Rourke's political compromise achieved was to close off a future energy supply option for Ireland.

Joe Jacob, who took over Emmet Stagg's nuclear brief as Minister of State for Energy in the new Fianna Fáil/Progressive Democrat coalition in 1997, could claim in all sincerity: 'There is no country on this planet that is more anti-nuclear than this one, no Government more anti-nuclear than this one and no people more anti-nuclear than the Irish people.'[3]

* * *

To BNFL's frustration, the company's application to operate the new Sellafield MOX plant (SMP), built at a cost of £300m in 1996, appeared doomed to endless rounds of public consultation. BNFL had applied to the Environment Agency for authorisation of discharges from SMP in 1996 and two rounds of public consultation, including the case for its economic justification, had already taken place by the end of 1998.

For more than thirty years experimental MOX fuel had been manufactured at Sellafield in B33, an old building on the site, by the UKAEA and BNFL for use in the fast-breeder reactor programme and research purposes. MOX fuel, with a typical content of about 5 per cent plutonium oxide, was first produced for commercial use in light water reactors throughout Europe from the early 1980s. Anti-nuclear campaigners focused public attention on the number of Nagasaki-type bombs that, theoretically, might be manufactured from its plutonium content, and security risks attached to plutonium transports.

From late 1993, BNFL started manufacturing MOX fuel for customers in Switzerland and Germany in the MOX demonstration facility, or MDF, which BNFL had constructed within B33. The Sellafield-manufactured MOX fuel was transported, first by air from Carlisle and latterly by sea using an industrial roll-on, roll-off ferry, the MV *Arneb*, when new IAEA specifications for fuel transport flasks effectively brought air transport to an end. BNFL did not seek renewal of an air transport contract with Carlisle airport that expired in early 1999.

At the end of 1997, MDF began to manufacture fuel to a Japanese design. Successful delivery of this first consignment was crucial to BNFL's long-term aim to secure Japanese contracts for its new MOX plant. Eight assemblies of MOX, destined for the nuclear reactors at Takahama power station, left the BNFL marine terminal at Barrow-in-Furness in July 1999 on board the purpose-built nuclear transport ship, the *Pacific Pintail*, accompanied by its sister ship, the *Pacific Teal*, acting as a security escort and a decoy, for the port of Fukui.

To BNFL, the delivery of this fuel represented the completion of the nuclear fuel cycle. MOX fuel for Japan was the final technical and commercial step towards securing the company's long-term future. For the anti-nuclear groups, especially Greenpeace, the voyage was a demonstration of the 'plutonium economy' against which they had been campaigning for over twenty years.

That same month, the Labour government announced that the sale of 49 per cent of BNFL's shares would take place before the next election. The part privatisation was expected to raise £1.5bn for the exchequer. Officials in the Department of Trade and Industry hinted the eventual sale of BNFL's remaining 51 per cent shares would follow in due course.[4]

With John Guinness's tenure as BNFL Chairman drawing to a close, Hugh Collum, whose thirty years' experience of financial management in the pharmaceuticals and energy sectors showed he was well-equipped to bring BNFL to the market, was nominated as his successor in October 1999. From a corporate perspective, BNFL enjoyed strong support at the centre of government. The company had an annual turnover of £1.4bn, although there remained doubts about whether it could meet pre-privatisation financial and business targets. Doubters in the city, among Blair's cabinet and even within the corporate ranks of BNFL itself, were outnumbered by optimists who believed the company was well on course to becoming a flagship British industry in the international arena.[5]

New employment contracts and procedures at Sellafield had resulted in employment levels at the site reducing by about 1,500 workers since 1994. The job reductions at Sellafield caused tension between BNFL's CEO, John Taylor, and Lawrence Williams, head of the Nuclear Installations Inspectorate. To many in BNFL, Williams' antipathy to their CEO was becoming personal.

An increase in the number and frequency of incidents reported in 1998 prompted the NII chief to mount yet another investigation into safety at Sellafield. A team of thirteen NII inspectors was already on site conducting a root and branch examination of Sellafield safety culture on Friday, 10 September 1999 when the UK *Independent* informed BNFL press office it was about to publish a story that quality assurance data on MOX pellets manufactured in MDF had been systematically falsified by workers at the plant.[6]

* * *

BNFL had a weekend in which to prepare for public disclosure of the scandal that would ultimately destroy the company. It was time to tell the NII and its Japanese fuel agent, Mitsubishi Industries, what the Sellafield management already knew – one worker had admitted to falsifying data; another had confirmed he knew what was going on. Unions and management at the plant had met to discuss the implications of the data falsification. As to its extent and duration, that was more difficult to pin down, because tens of thousands of data spreadsheets needed to be examined.

There were aspects to MOX manufacture in MDF so tedious they would hardly have been out of place in a Heath Robinson cartoon. Primary quality assurance checks on the size of individual MOX pellets were automated to discard any that were too large or too small, but secondary sample inspections had to be carried out as specified by the individual customer. In the fuel under manufacture in 1999, manual secondary checks, using a spectrometer and noting the individual measurements, were required on random lots of 200 pellets. On 20 August 1999 a shift supervisor in MDF noticed similarities between different sets of data on the spreadsheets. Shift workers had settled on the neat trick of copying results from old data sheets.

BNFL's executive directors immediately faced a difficult decision – to turn back the ships en route to Japan and weather a negative PR storm or allow the ships to continue on the assumption that the fuel bound for the Takahama reactor was untainted.

'There was an awful lot of pressure on Taylor to face it out and "stick with it",' Colin Duncan, then BNFL's Director of Communications, would later recall.

'In hindsight if it had been possible for him to turn those ships around, order a full enquiry and announce the closure of the MOX plant, I think he'd have kept his job,' Duncan said.

From a practical point of view, the problem was that the ships were north of Fiji, three-quarters of the way to Japan at the time. They did not have enough transport fuel to turn around and return to Britain and there was no contingency plan for getting fuel from elsewhere. In any case, an oil delivery from any nearby port might have been difficult to organise, given the nature of their cargo.

Alternatively, BNFL could offer to take the MOX back, if the Japanese utility, Kansai Electric, so requested. Tom McLaughlan, now stationed as science officer in the UK embassy in Tokyo, and Gavin Carter, who worked for BNFL Transport, suggested this approach to the Director of Reprocessing, Chris Loughlin.

'Loughlin angrily rejected the advice and instructed us not to mention the idea to anyone,' Carter later recalled.

On 21 September, BNFL issued a press statement that the Japanese fuel was unaffected by the data falsification. By mid-December BNFL found itself having to publicly confirm that four of the eight MOX assemblies landed in Japan were tainted. Within the company the argument again focused on whether BNFL should immediately offer to take the MOX back. The automated primary checks meant it would be safe in use in a reactor. The Director of Reprocessing, initially backed by DTI officials and ministers, favoured refusal. The Japanese proved just as obdurate in their insistence that they would not accept this fuel.

BNFL's Japanese customers and its government were deeply offended by the company's stance. BNFL, they said, would not receive any further contracts for reprocessing Japanese fuel or for MOX manufacture. The British government was seriously embarrassed, triggering a collapse of political confidence in BNFL senior management.

Apart from the UK *Independent*, which tantalised its readership with regular updates on the progress of events, the data falsification story remained a media slow-burner until 17 February 2000 when the

NII simultaneously published three reports: on the MDF scandal, the storage of HAL waste, and safety at Sellafield.

At corporate level, BNFL was aware of the potential impact of publication of these reports on its international reputation and had made detailed preparations to deal with the aftermath. As often seemed to happen, the public affairs team appreciation of what was coming down the line was out of synch with the BNFL press office interpretation. A few days before publication, BNFL press office adamantly insisted the reports would attract local publicity only; that this was not a national story. They anticipated, wrongly as it turned out, that a statement about data falsification in MDF going on since 1996 and also affecting MOX previously supplied to Switzerland and Germany would be withdrawn before the report was published. There was no crisis. BNFL's new chairman had gone skiing.

The data falsification report and the NII's damning indictment of safety culture at Sellafield provided the lead item for every news bulletin and newspapers in Britain and Ireland from early morning on 17 February and quickly developed into a major international story. UK Energy Minister, Helen Liddell, called for a 'root and branch' review of BNFL management the following day. In the US, the Department of Environment threatened to cancel clean-up contracts with BNFL's US subsidiary, BNFL Inc. European customers, including Germany, spoke of withdrawing reprocessing contracts from Thorp.

Frank Turvey was a member of the RPII team at Sellafield examining the safety case for HAL storage when the story broke. The RPII team was hastily withdrawn.

The Sellafield unions decided they could not defend the five shift workers at MDF who had already been sacked by the company. The ability of those workers to cut and paste was of a standard that should have won an IT award, one of the shop stewards remarked. The NCNI released a statement, bitterly opposed by the Director of Reprocessing and the Sellafield management, calling for the appointment of a 'safety tsar' and immediate postponement of BNFL's privatisation.

'In the midst of what was arguably the beginning of the end of BNFL there were people who still thought the issue was privatisation,' Jim Innes recalled. 'It showed me the British nuclear workforce can defeat any enemy except the British nuclear management.'

Under the unrelenting glare of continuing bad publicity, morale plummeted at Sellafield and within BNFL generally. John Taylor had lost his job by the end of February; the Director of Health and Safety, David Coulston, also left the company.

As executives struggled to make sense of the catastrophe unfolding around them, wild rumours began to circulate that the MDF saboteurs were paid agents of Greenpeace or even the CIA. A major internal investigation was launched to discover the source of the original leak to the UK *Independent*. Some executives at Risley began to speak resentfully of the capacity of the Sellafield management to hang together and support one another, irrespective of what was happening or how far they had fallen short in fulfilling their responsibilities, individually or collectively. Freemasonry, always pervasive among the professional classes in England's north-west, was the glue that held the Sellafield management together, people said. Others felt it was irrelevant whether or not Sellafield senior managers were members of a Freemasons lodge.

'They were engineers. And not only that; they predominantly came from the same tiny stratum of North of England society; geographically, socially, economically, politically,' one source said. 'They didn't need to be Masons because they were Masonic enough themselves already.'

From Colin Duncan's perspective, the isolationist mentality at Sellafield went beyond what would normally prevail between an operating site and company headquarters. Duncan was aware of the rumours about the influence of freemasonry at Sellafield, but like many other strange events that occurred during this period, it was just another thing that nobody ever quite got to the bottom of.

'Everybody knows everybody else,' Duncan described West Cumbrian society. 'There's a huge dependency on Sellafield economically and psychologically. And so that leads to a generally introspective mindset. Of course it needs to be the opposite really for the business that they're in.' 'I don't believe it will ever really change,' he added.

What did change was BNFL's relationship with government and its prospects for a privatised future. The MDF scandal ended the relative autonomy over its own business that BNFL had previously enjoyed. Privatisation was off the table, temporarily as far as some of BNFL's senior management were concerned; permanently in the view of most neutral observers.[7] As BNFL's financial position worsened, partly pulled down by the costs of operating the magnox stations and their inestimable liabilities, the DTI and the Treasury began to take more control over how the company's business was managed. At Sellafield, the NII assumed a much more directive role on safety and operational standards.

* * *

At first the Irish media, as one environment correspondent admitted, failed to realise the full significance of the MDF scandal, at least until the NII reports were published and the story gained momentum internationally. As elsewhere, the impact of the story in Ireland was to destroy BNFL's credibility, particularly on safety at Sellafield. An *Irish Times* editorial in 1997 had described Sellafield: '... to be fair to BNFL, as probably one of the best run nuclear plants in the world ...'[8] Now BNFL were branded as 'liars', and worse, in the Dáil and in the media generally.[9]

A peculiarly Irish twist to the tale arose from the theft from Rupert Wilcox Baker's computer of a large volume of internal correspondence going back over several years. Passed on to Channel 4's 'Dispatches', the correspondence formed the subject matter of a programme broadcast on 13 April 2000. The hacked material included letters and memos relating to a response to the Wicklow independent deputy, Mildred Fox, who had written to Tony Blair in 1997 to congratulate him on his election as Prime Minister and urge him to close down Sellafield.

Wilcox Baker referenced the Wicklow deputy's letter as 'election politicking' and drafted a reply for the UK Minister of State for Science, Energy and Industry, Mr John Battle. BNFL wrote separately to Ms Fox, extending an invitation to visit Sellafield that Ms Fox availed of that summer. Mildred Fox told Channel 4 she found BNFL's treatment of her letter to Blair 'very offensive and condescending'.

Privately, journalists would acknowledge there was nothing unusual in a semi-state organisation being advised of such correspondence by departments and asked to suggest an appropriate response. But dramatic headlines about 'furious diplomatic rows', and Bertie Ahern's promise to raise the matter with Tony Blair, made for better copy.

In Britain, disclosures on BNFL's rail transport company and suggestions on how various MPs should be handled generated even greater fury. Jo Moore, special adviser to DTI Secretary of State, Stephen Byers, later infamous for her own efforts on good days to bury bad news,[10] pressurised Colin Duncan to have Wilcox Baker dismissed from BNFL. Duncan resolutely refused.

* * *

The data falsification scandal prompted Joe Jacob to ask Ireland's new Attorney General, Michael McDowell, to once again examine Ireland's options for an international court case.[11]

Jacob was enjoying mixed success as the minister responsible for leading Ireland's fight against Sellafield. At an early stage of his ministerial office, he had been taken aside and advised at the highest level to make the battle political, rather than simply echoing the views of his civil servants and the RPII.

The issue at the core of the Louth Four damages claim – the health impact of Sellafield discharges, historically and from Thorp – was undermined by publication of Dr Geoffrey Dean's report in November 2000 on further research into the Dundalk Down's syndrome cluster.[12] Of the six girls identified in the original cluster as subsequently giving birth to children with Down's syndrome, three were shown not to have been pupils of the St Louis convent school in Dundalk or even living in the area at the time of the 1957 'flu epidemic and the Windscale fire. The North Eastern Health Board announced it had no plans for any further research into the causes of the cluster, and in April 2001, the RPII accepted Dean's findings as conclusive.

'The suggestion of a link between the Down's syndrome births and the Windscale fire is unfounded,' the RPII stated. 'For many years there has been a widespread belief that the existence of such a link was probable or even proven, and this belief has undoubtedly been a source of anxiety for people in Ireland, particularly in the Louth/ Dundalk area.'[13]

'The problem is that so many people don't want that answer,' Fergus O'Dowd admitted later. O'Dowd accepted that Sellafield did not cause the Dundalk Down's cluster, but he noted: 'People will always link cancer to Sellafield. I've met many, many people who believe that they got cancer from Sellafield.'

Under pressure from the opposition because of Fianna Fáil's election commitment to support the Louth Four case, the government had set aside £400,000 as a research budget for the plaintiffs' use, but avoided paying any bills that might subsequently be construed as legal fees.

Joe Jacob also embarked on a diplomatic coat-trailing mission to the Nordic countries to seek support for Ireland's campaign against Sellafield, specifically technetium 99 discharges. His efforts appeared to have been rewarded in July 1998 when all fourteen OSPAR member countries and the EU signed up to the Sintra Statement in Lisbon, a commitment to further reduce radioactive discharges to the marine environment by 2020, that was partially based on Irish proposals. BNFL and the British government had no difficulty with the OSPAR commitments, well aware as they were that Sellafield could meet its Sintra obligations when magnox reprocessing came to an end as it was scheduled to do by 2012, and without affecting Thorp reprocessing.[14] Jacob's understandable pleasure in the outcome of the OSPAR Summit was marred by Greenpeace criticism that the Irish delegation in Lisbon had not gone far enough in pressing for a complete end to reprocessing at Sellafield.

Like every Irish minister before him, Jacob could never overlook an opportunity to raise Sellafield internationally. Brian Wilson, the Scottish MP appointed UK Energy Minister in early 2001, later recalled an EU meeting at which Jacob, who had been quiet for most of the meeting, unexpectedly made an 'out of context' intervention about Sellafield.

'There was a flurry of activity between the officials and it was agreed that we would have a private meeting in the margins,' Wilson remembered.

'We went outside to the coffee bar and we spoke about fishing, we spoke about the Irish language, we spoke about hurling, we spoke about the weather,' Wilson said. 'The last thing it seemed to me the Irish Minister wanted to talk about was Sellafield because he obviously had no knowledge or understanding of any of it.'

* * *

The Irish government wanted access to financial details contained in a report on the commercial viability of the Sellafield MOX plant prepared by the accountancy firm, Arthur D. Little.[15] Investment and commissioning costs in SMP had risen to £470m without one MOX pellet being manufactured in the plant and the project was now on its fifth round of public consultation. Ireland's request was refused on grounds of 'commercial confidentiality'.

Although they were familiar with Irish demands for further information during project consultations, neither the British government nor BNFL could countenance releasing contract details that might inevitably find their way to their French competitors for MOX business. Persistent refusal to give access to the financial information led to Ireland requesting an arbitration hearing at OSPAR in May 2001 seeking 'provisional measures', the international equivalent of a local interim injunction, that they hoped would place moral and political pressure on the British to release the confidential data on MOX pricing. The Irish government requested the UK to make no decision authorising full production at the MOX plant while these arbitration proceedings were in progress.

Early in the afternoon of 11 September 2001, a group of BNFL managers at Risley sat down to plan the communications strategy for the return of the MDF-tainted fuel from Japan, scheduled for the following year. The meeting quickly adjourned to watch the fall of Manhattan's Twin Towers as the tragedy unfolded on television.

Three weeks later on 3 October, Secretary of State for the Environment Margaret Beckett announced the go-ahead for the Sellafield MOX plant. Inexplicably, no formal diplomatic arrangements were in place to advise the Irish government of the decision in advance of its public announcement, nor did the British embassy in Dublin attempt to make any contact with Irish officials. BNFL expected the usual pre-notification procedures would apply and once the announcement was made in London dispatched letters to the minister, Department of Energy officials and other political contacts in Dublin. It was the first many of them were to learn of it. As Taoiseach Bertie Ahern later remarked resentfully at a Fianna Fáil party conference, 'We didn't even receive the courtesy of a phone call.'

In London, Greenpeace and Friends of the Earth lodged papers in the High Court seeking a judicial review. In the Dáil, the Taoiseach promised he would raise the issue with Tony Blair and consider further legal action against Britain.[16] On RTE's current affairs programme, 'Prime Time', Joe Jacob promised: 'We will continue with our case under the OSPAR arbitration process; we will take the main case under EU law, and we will win, because we must win this.'[17]

* * *

Jacob had made a presentation to a cabinet meeting in June 2001 setting out Ireland's options for an international court case against Sellafield. The Irish government could seek arbitration under the OSPAR Convention, or a ruling from an UNCLOS Tribunal under the International Law of the Sea, or take a case to the European Court of Justice under Directives 142 and 227 of the Euratom Treaty.

Attorney General Michael McDowell made a further presentation to the cabinet meeting on the morning of 25 October 2001. Later that day the government announced it was seeking an immediate injunction from an UNCLOS Tribunal to halt the operation of the Sellafield MOX plant.

The move wrong-footed the opposition: a motion from Labour's Emmet Stagg setting out a ten-point action plan to close Sellafield, including international court action in the EU, was being debated in the Dáil.[18] The Labour Party had staged a 'Shut Sellafield' demonstration at the gates of the British embassy on Merrion Road and the party leader, Ruairi Quinn, handed in a letter of protest to the British ambassador.

While it was little understood in Whitehall, the priority for Bertie Ahern's administration was to leave no hostages to fortune, or the opposition, in the forthcoming 2002 general election. As the first Irish government to take international legal action against the UK over Sellafield, Fianna Fáil's political position was unassailable. Minor digressions from the commitment to wholly fund the Louth Four court case would be forgotten and forgiven by the electorate. Nor was any question raised, or explanation offered, as to why, after twenty years of stating the opposite, the views of the Attorney General's office on the sustainability of an international legal challenge had now changed.

* * *

Despite its reputation for secrecy, BNFL was accustomed to placing a large volume of information in the public domain, not least because regulatory requirements demanded publication of monitoring results, and any site incidents with a radiological dimension had to be reported.

Almost constantly in the media spotlight, BNFL published extensive information on Sellafield and site activities as a matter of course. More was extracted in NGOs' High Court cases, a regular event in the company's never-ending battle with Greenpeace and Friends of the Earth, for whom Thorp retained its iconic emotional status, now expressed in their opposition to the Sellafield MOX plant. BNFL was also engaged in a national 'stakeholder dialogue' in Britain at which NGOs, including Greenpeace in the initial stages, sat down with the company representatives and other interest groups in an attempt to reach consensus on the major issues facing the nuclear industry.[19]

'BNFL were a lot less secretive than the oil industry,' Colin Duncan observed. 'Or the French nuclear industry, come to that.'

The script was changed again by 9/11, which brought the threat of terrorist attack on nuclear installations to the forefront of public and political concerns. The security regulator for the nuclear industry, the Office of Civil Nuclear Security (OCNS), had its origins in the UKAEA nuclear constabulary. The OCNS was moved into the Department of Trade and Industry in October 2000. In pre-9/11 days, OCNS inspectors often visited Britain's civil nuclear sites 'clipboard in hand'[20] to assess security requirements and suggest site improvements. The imminent 'war on terror' led to procedures being streamlined and a more powerful role for the OCNS in directing the British nuclear industry's public presentations.

BNFL undertook a reassessment of the security of its assets on the Sellafield site in the wake of the 9/11 attack. Safety case assessments for sensitive buildings at Sellafield, including the HAL facility, Thorp and plutonium stores already took the possibility of accidents involving light aircraft or helicopters into account. More than forty scenarios including missile and aircraft attacks on the site were now assessed in conjunction with MI5 and MI6.

Most BNFL plants were not specifically designed against aircraft impact. For some newer and high-hazard plants, the structural design addressed the direct impact of a limited range of aircraft, but not the effects of fuel fires. BNFL was well aware it would be difficult to make a credible case for resistance to a deliberate crash of a fully fuelled airliner into most of the Sellafield facilities.

Many of the more recent plants built at Sellafield were designed to withstand seismic events and were considered less vulnerable to any major release of radioactivity following a deliberate crash by a large airliner. Thorp, for example, was judged to be practically invulnerable, with the exception of its storage pond area. However, older structures on the site, such as the B30 pond, the control rooms in the HAL facility, or the waste storage silos at B38 and B41 were not. Nor were the Calder Hall reactors, whose cooling pipes, brightly painted, stretched across the exterior walls of the reactor buildings.

Some protection was afforded to these older structures by the very layout of the Sellafield site, its overall congestion and the presence of tall chimneys and the Calder Hall cooling towers, for instance. The thinking was that site congestion and obstacles like stacks and towers would obstruct the flight path of any aircraft and make targeted pinpointing of any particular building on the site well-nigh impossible.

The threat of terrorist attacks on nuclear sites, or of terrorists' securing access to nuclear materials, had always ranked high among campaigners' objections to new plants at Sellafield, including Thorp and the MOX plant. Now, the terror threat scenario went into overdrive. Greenpeace set about commissioning a series of reports that would advance the most extreme consequences of a successful aircraft attack on Sellafield's HAL facility, its plutonium stores, B30 and a host of other installations on the site. Simultaneously, the WISE Paris group, headed by the Franco-German campaigner, Mycle Schneider, was working on a report on reprocessing at Sellafield and Cap de la Hague for the European Parliament scientific committee, STOA. The *Irish Independent*, under the heading 'Sellafield terror attack "would wipe out 3.5m"', would subsequently claim that 'the consequences of such an attack – in the case of Sellafield only sixty miles from Ireland – is so alarming that Greenpeace is unwilling to publish the reports.' Greenpeace had 'sat on' the reports, prepared by former Aldermaston scientist, Dr Frank Barnaby, engineering consultant, Dr John Large and the WISE Paris organisation, 'unable to decide what to do with them'. The British government, the paper claimed, had 'sought to suppress' the Greenpeace findings.[21] Despite the rhetoric, at the time copies of these studies were freely available to download from the Oxford Research Group (ORG) website.

The terrorist threat was being highjacked by anti-nuclear groups who 'appear to be using the events of 11th September for their own ideological aims', BNFL said.[22] But after years of open communications and public engagement, BNFL public affairs staff and Sellafield managers found they were faced with new constraints on what they could and could not say in public. Information that had previously been freely available in the public domain and in BNFL publications and on the company's website was withdrawn. An atmosphere of self-censorship set in among the company's executives. Data that had been openly presented to the public about standard safety issues at Sellafield in the pre-9/11 era were excised from presentations and statements.

Fergus O'Dowd, now a Fine Gael senator, had invited BNFL to make a presentation at a public meeting he planned to hold in Drogheda on 3 November 2001. The theme of this meeting was safety at Sellafield in the face of the new threat of terrorism. Suddenly, for the first time ever in a decade of campaigning in Ireland, there was a question mark over whether the BNFL delegation would be allowed to attend the Drogheda forum.

21

Pyrrhic Victories

'I know the British don't like it, but I get on well with Prime Minister Blair on other issues. I know our campaign on [Sellafield] irritates him, irritates his government and irritates the British, but we have made it absolutely clear that we're committed to this locally, in Europe and internationally. And you know, there's only one way we'll be satisfied in the end and that's when the British make a strategic change away from nuclear.'

Taoiseach Bertie Ahern, 7 September 2004

Joe Jacob didn't want to give rise to 'alarmistic vibrations' about the consequences for Ireland of a terrorist attack on Sellafield, he told RTE in a radio interview on 26 September 2001.

'We mustn't be alarmistic. We must be reassuring,' the minister implored as he fended off repeated questions from broadcaster Marian Finucane as to when she would get her iodine tablets.[1]

The net result of this pythonesque interview was a government decision to distribute potassium iodate tablets to every household in the country, at a cost of over two million euro, and the end of any prospects for further advancement in Joe Jacob's ministerial career, as the tabloid press bayed for his resignation and Taoiseach Bertie Ahern was photographed with his arm around his beleaguered minister's shoulders, announcing that he would stand by him.

Under the headline 'Target Sellafield', the *Star* newspaper reported that CIA chief George Tenet had warned the UK authorities Sellafield was on the Al Qaeda target list, although no such briefing ever took place.[2] Another paper, *Ireland on Sunday*, claimed Osama bin Laden was heading up the Irish Sea towards Sellafield in a submarine.

Ireland provided fertile ground for scaremongering. The WISE Paris organisation, already under contract for a report on European reprocessing to the European Parliament Scientific Committee, STOA, hastily repackaged data from their main report into a post-9/11 assessment of the likely result of an airline crash onto Europe's reprocessing facilities at Sellafield or Cap de la Hague and leaked the results. A million deaths beckoned. The radioactive inventory of the HAL tanks at Sellafield was over a hundred times what had been released in the Chernobyl accident, Mycle Schneider of WISE Paris later told the 'Marian Finucane Show'.

'Even a fraction of that radioactivity being released would lead to a disaster which would be beyond imagination,' he asserted.[3]

Attempting to inject some common sense into the mounting hysteria, the RPII CEO, Dr Tom O'Flaherty, took to the airwaves to explain some basic facts.

'[Sellafield] is one hundred miles away and not sixty as is often said,' Dr O'Flaherty said.[4] Given the distance, nobody in Ireland would suffer any immediate perceptible health effects from an accident or attack on the site and in any case it would take several hours, if not days, for any radiation released to reach Ireland.

'The concern would be that there will be exposure to radiation, which would increase the risk of cancer in future years. So the important thing would be to minimise the exposure so as to minimise that risk,'[5] O'Flaherty explained. The most effective means of doing that, as his own minister had tried in vain to impress on Marian Finucane, was to take shelter indoors.

Speaking to RTE news, the Minister for Defence, Michael Smith, chair of the government task force to assess intelligence relating to the possibility of any terrorist attack on Ireland, categorically rebutted the suggestion of a terrorist attack on Sellafield. His committee, he noted, had no knowledge of any such 'credible threat'.[6]

But to a public fed for decades on a diet of political propaganda about the dangers of Sellafield to Irish lives and the environment, 'no threat' reassurances from politicians were what appeared incredible.

* * *

The real damage to Ireland from a terrorist attack on Sellafield was likely to be more psychological than radiological. Fear of Sellafield was so ingrained in the popular imagination that widespread panic in Ireland would most likely follow reports of any terrorist attack or major incident at the site. Official reassurances of no danger to Irish east coast communities could hardly be expected to carry much weight at home. Abroad, Ireland's agricultural and high value-added seafood exports and its tourist industry were at risk of severe and possibly long-lasting damage. Risk perception, it was turning out, was more important than the actual risk itself.

At Fergus O'Dowd's Drogheda conference on 3 November 2001, Dr O'Flaherty set out the RPII's expert opinion that a major accident at Sellafield would produce no immediate casualties in Ireland. The Calder Hall power station and the HAL facility were the two main installations at Sellafield that, in the event of an accident or terrorist attack, had the potential to release enough radiation to impact on Ireland. The main risk from the Calder Hall reactors was a release of iodine 131, but Calder Hall was scheduled to finally close by 2003. The HAL tanks contained no iodine 131 and the principle radionuclide likely to be released was caesium 137, familiar to most Irish audiences as the main contaminant in Sellafield discharges to the Irish Sea and in the fallout from Chernobyl that still affected sheep in the uplands, fifteen years after the accident.

The RPII's report on its examination of the HAL safety case had already concluded that any major accident at that facility was unlikely.[7] Following a terrorist attack on Sellafield that might conceivably result in a release of radioactivity, Dr O'Flaherty stressed, the extra number of deaths from cancer in Ireland over a fifty-year period would be small and in any case virtually undetectable against the background of all cancer deaths. Such eventual fatalities would be directly related to individual radiation exposure, provided, of course, that contamination from the incident ever reached the country. On the day, no great enthusiasm or relief for this measured analysis was discernible among the Drogheda audience.

Scenarios predicting that the entire radioactive inventory of the HAL tanks could be released in some sort of domino effect if one of the tanks was breached by a large aircraft followed by a major fuel fire were implausible, since the tanks were effectively sealed off from one another, the RPII would persist. As late as 2005, at a meeting with the Joint Committee on Environment, the RPII's David Pollard reiterated their expert view: 'I cannot foresee any mechanism whereby one tank would impact on the next,' he said.

'The only scenario in which more than one tank could be impacted would be in the case of a catastrophic event which affected the cooling system of several tanks,' he suggested. In Pollard's opinion this was unlikely. Since it would take several days for the highly active liquid in individual tanks to reach boiling point, there was ample time for human intervention to prevent any major radioactive releases.[8]

BNFL had already calculated a fully fuelled aircraft crashing onto the site might not result in any major release of radioactivity, or at least any that would go far beyond the site boundary or its 20 km emergency zone. But a raging fuel fire might cause up to 4,000 casualties on the site if, for example, the plane struck the main administration building. BNFL had already upgraded fire fighting equipment at Sellafield to airport standards and was considering moving its administration block to a different location. Reinforced concrete walls to break the path of any incoming aircraft were under construction around its plutonium stores.

With genuine regret on both sides, the informal contacts between the RPII and BNFL that had facilitated the flow of information, and provided reassurance about the significance of various events and incidents at Sellafield over many years, was ended. Both Rupert Wilcox Baker and John Clarke, then Safety and Environment Manager at Sellafield, came under pressure within BNFL, from their regulator and from the DTI, for extending an invitation to the RPII at the Drogheda conference to review enhanced safety arrangements for HAL storage.[9] Within months, the RPII were advised that from now on they must rely on their UK counterparts in the NII for information about developments at Sellafield.[10] Any discussion on site security was entirely off the agenda. What might previously have been viewed as 'safety' issues were now increasingly merged with 'security', adding to the RPII's difficulties in making safety assessments of Sellafield installations.[11]

* * *

Officials in the DTI and the NII had become increasingly impatient at the way in which Irish politicians were cranking up the pressure on the terrorism issue. The politicians were providing plenty of examples for recalcitrant British officials to draw on.

Over lunch in Drogheda, Fergus O'Dowd had informed Wilcox Baker he proposed to visit Sellafield the following Tuesday, 6 November 2001, to make a programme on site security with a Channel 4 TV crew. It was the first Wilcox Baker learned of this adventure. The Sellafield press office had turned down the Channel 4 request the previous week, as an RTE film crew was already scheduled to visit the site on that date.

His request could not be accommodated at such short notice 'at a time of heightened security', Wilcox Baker wrote to O'Dowd the following Monday.

'To hear at one working day's notice, that you intend to visit Sellafield at the invitation of a third party we consider is not an appropriate way to arrange such an important visit and one which has been the subject of such a longstanding invitation from BNFL,' he stated.

'I couldn't understand why they wouldn't let us in. I still don't,' Fergus O'Dowd would insist several years later. 'I flew over there. We drove overnight. It was a very successful trip. It brought unbelievable attention to the issue,' he said.

'I got more mileage out of that than anything else in my life. It put the issue of Sellafield back on the agenda again nationally.'

Not to be outdone, an advertisement calling for the closure of Sellafield, signed by all members of the Fianna Fáil Parliamentary Party, including the Taoiseach, Bertie Ahern, was published in the *Sunday Times* on 24 November 2001. The ad, Fianna Fáil sources said, was paid for from the left-over budget of exchequer support for political party activities. A few days later, Ireland's case seeking an interim order to halt commissioning of the MOX plant got under way before a special ITLOS Tribunal in Hamburg.

On 4 December 2001, the new Northern Ireland Assembly unanimously passed a motion demanding the closure of Sellafield. It was ridiculous, the Rev. Dr Ian Paisley stated, that the minister with responsibility to the people of Northern Ireland on Sellafield had not been consulted about the go-ahead for the Sellafield MOX plant. 'The British Government must be called to account,' Dr Paisley warned.[12]

* * *

An increase in radioactive discharges to the Irish Sea from the Sellafield MOX plant was central to Ireland's international UN case. But SMP was mostly a dry-process operation. BNFL had initially considered incorporating SMP's practically negligible discharges into its application for Thorp to the Environment Agency.

Ironically, Ireland was taking a case based mainly on Sellafield discharges at a point where their impact on Ireland was at its lowest in over twenty years. The main pathway for exposure to Sellafield radiation was through the consumption of seafood landed at ports on Ireland's east coast.

The annual dose to a heavy seafood consumer had fallen from 0.160 millisieverts (mSv), that is, 160 unit doses of radiation, in 1976 to less that 0.002 mSv, or two units, by 2001, and to a 'typical' seafood consumer from 0.030 mSv to 0.0003 mSv.[13] Naturally occurring polonium 210 in fish and shellfish delivered a much higher radiation dose than Sellafield discharges: an annual dose to the heavy consumer of 0.150 mSv, or 0.030 mSv to the typical consumer. Even at their peak in the mid-1970s, the impact of Sellafield discharges on the human health of the Irish population would hardly have given rise to serious concern.

The average radiation dose to members of the Irish public, from all sources – medical, solar, radon from rocks and soil – is about 3 mSv, or 3,000 units, per annum. The ICRP recommended dose limit for members of the public, in addition to naturally occurring background radiation, is 1 mSv. The RPII estimated the risk to a heavy seafood consumer of developing a fatal cancer at about one in thirteen million and the risk to a typical consumer at one in sixty million.[14]

The European Commission's 2002 Marina II study showed that North Sea oil and gas operations were contributing more man-made radioactivity to north European marine waters than the nuclear industry. Naturally occurring radioactive materials (NORM)[15] had become the dominant source of radioactive doses to the population of the European Union (EU) from industrial discharges ahead of discharges from Sellafield or Cap de la Hague.[16]

On 3 December 2001, the ITLOS Tribunal announced it was rejecting Ireland's request for preliminary measures to stop plutonium commissioning of SMP. The tribunal, however, ordered both countries to enter discussions with a view to resolving their differences and submit progress reports on their consultations every six months.[17] The opportunity to snatch a PR victory from the jaws of defeat was not wasted. The tribunal's ruling was a victory for Ireland, Attorney General Michael McDowell insisted.

'For the first time, Ireland has an international ruling that accepts that Ireland has an interest in Irish Sea pollution levels,' he said. The glass was half full, not half empty, and the substantive cases under OSPAR and the UN Law of the Sea would be heard in The Hague, some time in 2002. There was no need to state the obvious: politically, the Sellafield issue was now safely parked until after the next general election.

* * *

Much against the advice of his own officials in London, but with the enthusiastic support of the British embassy in Dublin, Energy Minister Brian Wilson had decided to take on the Sellafield debate in Ireland.

'I know Ireland quite well, and I was not prepared to be ill characterised in Ireland as some kind of uncaring monster – you know, the Brits sending nuclear waste across to Ireland,' Wilson explained. 'I was not going to have that pinned on me without challenging it.'

'Secondly the arguments were just intellectually insulting,' he added. 'The idea that they should be given free rein as they had been, by and large, by the British government for years I just thought was wrong.'

Wilson targeted what he regarded as the hypocrisy of Irish political opposition to Sellafield, such as the Fianna Fáil *Sunday Times* advertisement. He challenged the prevailing view in Ireland that the British government had less interest in protecting its own public and nuclear communities from the harmful effects of ionising radiation, or terrorist attacks that might lead to radiation release, than the Irish. Proper dialogue and shared information would 'be much healthier than a long procession of unfounded scare stories,' he suggested.[18]

Wilson was angered by the persistence of Irish claims, contrary to all the evidence, that Down's syndrome cases in Ireland were linked to Sellafield radiation. It was a poignant issue for him personally, as the father of a Down's syndrome child, and he was genuinely perturbed by what he perceived as the cynical manipulation of the radiation health issue in Ireland.

'All I am asking is that the discussion proceeds on the basis of scientific information and not on either emotive talk about what Sellafield is responsible for when it isn't, or else totally impossible demands about closing Sellafield,' Wilson pleaded in a radio interview. 'If we can get over the hump of ... sloganeering, and maximise co-operation and information sharing, then we can have the relationship of a critical friend rather than through tribunals and court hearings and rhetoric and wild allegations about what Sellafield does. That is all I want. To move it onto a sensible basis of debate.'[19]

'What also really offended me,' he noted later 'was the dishonesty of the "close it down" demand since, whatever else you might do about Sellafield, the one think you definitely couldn't do (not least in Ireland's own safety interests) was close it down!'

Brian Wilson came from a nuclear community – the Hunterston B nuclear power station was located in his constituency. As Energy Minister, he had announced the end of reprocessing at Dounreay, introduced the package to save British Energy from collapse, produced the February 2003 White Paper on Energy Policy and sponsored the policy that would lead to the establishment of the Nuclear Decommissioning Authority (NDA) to take ownership of Sellafield and BNFL's magnox sites in 2005, leading to the break-up of the company and the inevitability of its final dissolution.

Britain's electricity market privatisation had precipitated the 'dash for gas' of the 1990s, as the newest and cheapest form of electricity generation. By the end of the decade, as oil prices softened and gas prices fell even further, British Energy's previous £281m profits had turned to a £281m loss and the nuclear generator was in financial freefall. British Energy's fortunes would only recover when oil prices strengthened and wholesale electricity prices increased, but the folly of over-reliance on gas for electricity was also becoming apparent.

Domestic prices for electricity in Britain were by now 20 per cent above the average in France, despite the massively increased contribution of cheap gas generation. By 2020 it was estimated that Britain would rely on gas for 70 per cent of its energy supply, but with national reserves exhausted, most of this gas would be imported, leading to a potential crisis in security of supply for the first time in forty years.

Withdrawal of the older magnox stations from service and the progressive decline in nuclear's contribution to the British energy mix threatened to destabilise the energy market further. It was also realised that the loss of the nuclear supply – down to zero by 2020 if there was no new build – would contribute to Britain's problems to meet its climate change commitments.

In the 2002 energy review, Brian Wilson fought to keep the door open on the nuclear option.

'Given that billions were being forked out to bail out British Energy it wasn't the time to argue for nuclear energy,' he recalled. 'There were seven or eight ministers involved in the White Paper team and with the exception of myself and perhaps one other minister, nobody else was pro nuclear.'

The 2002 White Paper plumped for renewables as the solution to Britain's energy problems, setting an ambitious target of 20 per cent renewables supply by 2020, with extensive government support.

Within two years, it was clear that renewables could at most provide 4 per cent of Britain's energy supply by 2010, well short of government targets. As the conjunction between security of supply, rapidly increasing prices for oil and gas and climate change issues began to impact on energy politics, the government initiated a further review. This time nuclear took centre stage.

* * *

The first public confirmation that BNFL was technically bankrupt came in a *Sunday Telegraph* interview given by Philip Dewhurst, BNFL Communications Director, on 21 October 2002. Dewhurst told the *Telegraph* and the following day's *London Evening Standard* that BNFL was technically insolvent, but was not going bankrupt. The Chairman and Chief Executive of BNFL had proposed the creation of a Liabilities Management Authority to government, he said.

A review of its liabilities on the Sellafield site together with the likely decommissioning costs of the old magnox stations, the last of which, at Wylfa, was expected to close by 2010, had pushed company liabilities ahead of net asset values by 2001. BNFL chairman, Hugh Collum, and Chief Executive, Norman Askew, proposed the government undertake responsibility for the magnox liabilities. Declaring bankruptcy was their only other alternative.

The liabilities authority proposition was attractive to their government, but probably not quite in the manner assumed by BNFL's top management. All BNFL's liabilities would be taken over, and a new body, with a specific brief to accelerate clean up and nuclear decommissioning, would take ownership of BNFL assets, including those at Sellafield. BNFL would be broken up, and its key international assets, including Westinghouse and the US subsidiary, BNFL Inc., sold. A new BNFL subsidiary, British Nuclear Group (BNG), was established to operate as a contractor to the proposed Nuclear Decommissioning Authority on BNFL's old British sites. British Nuclear Group Sellafield became the operating company for the management of the Sellafield site.

Tony Blair's government was determined to get to grips with the historic waste problem, particularly at Sellafield. Unless and until it could demonstrate progress on dealing with the waste legacy, there was little prospect that Labour MPs, many of whom remained hostile to nuclear, never mind the public, might ever be convinced of the rationale for a new build programme. That scepticism was shared by what the UK *Independent* described as 'some well-placed members of the Government, including Margaret Beckett, the Environment Secretary, and Patricia Hewitt, the Health Secretary'.

'Even Alan Johnson, the new Energy Secretary, has expressed caution about nuclear power', the paper said.[20]

The legislation establishing the NDA made provision for continued operation of BNFL's former commercial assets, including Thorp, the Sellafield MOX plant, which was showing all the signs of a plant that didn't work, and the lucrative fuels transport business, to offset the costs to the taxpayer of the accelerated clean-up programme. The

NDA took formal control of Sellafield and BNFL's UK assets and sites in April 2005.

BNFL now effectively lost control of the strategic direction of their company to the Treasury, whose mandarins, and not the BNFL board, made the decisions about the disposal of the company's remaining assets, such as Westinghouse, BNFL's holding in Urenco, the European fuel services company, and ultimately the fate of British Nuclear Group itself. BNFL was reduced to rubber-stamping decisions taken elsewhere and the company's announcements on its future development sounded increasingly hollow.

Any fantasies entertained by British Nuclear Group of morphing into a world-class nuclear clean-up company, using the Sellafield contract as a launch pad and attracting other world-class engineering companies as partners in this endeavour, were dashed by events. Most notable was an accident in Thorp, discovered in April 2005, in which 83,000 litres of highly radioactive liquid leaked into a secure containment cell in the plant without being noticed for the best part of a year. The Thorp disaster, the *Independent* reported, could be the 'final straw for BNFL bosses' in their already fraught relationship with the government.[21] It was also a final straw for the NDA, which fined British Nuclear Group Sellafield £1m and made it apparent that the group would not retain the contract for running Sellafield when it came up for renewal in 2008.

Thorp was shut down. In January 2007 the NDA and the regulator, NII, granted permission for the plant to re-open once repairs were completed to the NII's satisfaction.

With Thorp expected to cease operations by about 2012, British Nuclear Group had been initially instructed by the NDA not to bid for any further reprocessing contracts for the plant. But the loss of income from Thorp increased the NDA's reliance on the exchequer to fund the clean up programme at Sellafield. By the close of 2006, BNG's commercial division was being exhorted to seek transport and new reprocessing contracts wherever they might find them.

British Nuclear Group was destined to be broken up and sold. It was only a matter of time before the British government would carry out its unstated intention and finally wind up BNFL.

* * *

As these profound changes in the structure of Britain's nuclear industry unfolded in the early years of the twenty-first century, they barely elicited even a passing comment in the Irish media or among the political classes. Instead, the Irish media and political focus appeared stuck in a late twentieth-century anti-BNFL time warp.

Throughout the spring of 2002 the Irish media was gripped by a 'Shut Sellafield' postcards campaign. Former advisor to the Labour Party leader, Dick Spring, Fergus Finlay, was the Svengali-like figure behind this campaign to send over a million postcards to Tony Blair, BNFL's Chief Executive, Norman Askew and Prince Charles. The campaign was spearheaded by Ali Hewson, wife of U2's Bono, a member of Greenpeace and patron of the Chernobyl Children's Charity run by her friend, Adi Roche.

The inclusion of Prince Charles on the target list was controversial and is believed to have cost Hewson the full endorsement of the Irish government for a truly populist campaign that shimmered and glittered with Irish celebrity endorsements, including the national rugby and soccer teams. It was also believed to have dissuaded RTE from broadcasting free advertisements to the campaign, on the grounds that the national broadcaster station was legally prohibited from running advertisements in support of a political cause – a matter of some disappointment to Ms Hewson.

But An Post undertook to deliver the postcards to every household in Ireland for free. Supermarket chains, though the British-owned Tesco stores proved the exception, stocked cards at their sales counters, and schools and universities displayed posters and held special sales days to boost the public uptake. Any profits were to go to Adi Roche's nationally popular charity, Chernobyl Children, which every year brought children from Belarus to Ireland for holidays with Irish families and delivered substantial aid to families and orphanages within the country that had suffered most from the fallout of the Chernobyl accident.

As the tabloid *Ireland on Sunday* would note at the end of the campaign, its funding mechanisms were not 'transparent' and 'remained a mystery'. The paper reported the organisers insistence that a €100,000 state grant to the Chernobyl's Children's Charity had not been diverted to the campaign, and that the government had, in any case, donated this money on the strict understanding that it could not be used for campaigning against Sellafield.

Adi Roche stood as a candidate for the presidency in 1997, for Labour, Democratic Left and the Green parties, but with some 6 per cent of the vote had only managed a poor fourth out of five in that contest. She could trace her anti-nuclear credentials back to the

Carnsore protests of the late 1970s and had been appointed by Labour's Emmet Stagg to the board of the RPII. Ms Roche was, and remained firmly convinced, irrespective of the emerging scientific evidence to the contrary,[22] that all disease and illnesses experienced in Belarus, especially to children, were a direct consequence of radiation poisoning from the Chernobyl accident. Sellafield, she regularly declaimed, was 'the Chernobyl on our own doorstep'.

Officials in the British embassy were initially keen to persuade BNFL senior management to meet with the campaigners, especially Ali Hewson, to discuss her concerns about Sellafield. Within the company other counsel prevailed: the postcards campaign enjoyed extensive media coverage in Ireland, albeit mainly confined to the tabloids, especially the *Star*, which ran 'a celebrity a day' spot in its support, and local radio stations. Despite a major PR drive, the campaign was almost entirely ignored by the British media. BNFL was not in the business of creating publicity opportunities for celebrities.

The campaign culminated on 26 April 2002 with Ms Hewson hand-delivering her own gigantic postcard to No. 10 Downing Street: 'Look me in the eye, Tony and tell me it's safe.'

Although the organisers remained adamant that well in excess of one million postcards were delivered, answering questions in the House of Commons, the Prime Minister, Tony Blair, claimed to have received only 187,000 letters and cards concerning Sellafield since the beginning of 2001.[23] The number sent to Prince Charles remained unknown, but from the weight of the bags on the postroom floor, BNFL estimated somewhere in the region of 300,000 cards had been posted to its own Chief Executive. The main issue preoccupying BNFL was at what point it would seem reasonable to send the contents of the post bags for recycling.

* * *

Later that year, in September 2002, the Pacific Nuclear Transport Limited (PNTL) ships carrying the tainted MOX fuel returned from Japan to Barrow-in-Furness. On their journey through the Irish Sea, the ships played cat and mouse with a Greenpeace flotilla of small boats, yachts and the Greenpeace *Rainbow Warrior*, based out of Dublin, that stalked their progress.

For BNFL, the return of the fuel was meant to bring closure to the MDF episode that had brought the company to its knees. Its return arguably also fuelled the last high-octane Irish protest against Sellafield.

Pyrrhic Victories

The Taoiseach, Bertie Ahern, his Environment Minister, and the Fine Gael leader, Enda Kenny, respectively, paid courtesy visits to the Greenpeace ship *Rainbow Warrior*, docked at Sir John Rogerson's Quay in Dublin.

'If Sellafield is hit by terrorists then death will be the least we have to fear,' Kenny stated. 'We've crept around BNFL for too long. It's a matter of fact – not opinion – that BNFL cannot be trusted.' Objections by the British embassy to the hysterical tone of this statement were brushed aside by the Fine Gael leadership.

The political and media hoopla that surrounded the BNFL fuel return in September 2002 was in marked contrast to a more discreet political silence that greeted the transport of 150 kg of US military plutonium in the *Pacific Teal* and *Pacific Pintail* from Charleston in South Carolina to Cherbourg in France almost exactly two years later in September 2004.

The US–Russia Plutonium Disposition Agreement, announced by President Clinton of the United States and President Putin of the Russian Federation on 4 June 2000, committed both countries to each disposing of thirty-four tonnes of weapons-grade plutonium over a twenty year period. The US military plutonium en route from Charleston was a first-stage implementation of this commitment. The plutonium was bound for the Areva-owned Cadarache facility in France for conversion into MOX fuel assemblies for return to the US. European Union support for the US and Russian anti-proliferation initiative, and Ireland's membership of the G8-led Global Partnership, left the Irish government with little room for manoeuvre. In any case, as one British official expressed it at the time, 'The Irish are terrified of the Americans.'

* * *

The possibility of a formal bi-lateral agreement between Britain and Ireland on nuclear matters, which BNFL had sought for ten years, was already under discussion among British officials and diplomats before Ireland's case finally reached the ITLOS Tribunal in The Hague in 2003.

The main obstacle to progress, as far as the British were concerned, was the persistent tendency among Irish politicians to whip up Ireland's battle against Sellafield into a national cause celebre. In June 2003, the Irish government awaited the outcome of its action under OSPAR. The hearing of its substantive case under the UN Law of the

Sea was set to commence in The Hague later that month. The two cases were described by then Minister for the Environment, Martin Cullen, as 'among the most significant legal actions ever taken by Ireland'.

On the British side, there was general confidence that Britain would successfully defend both actions. Britain was prepared to argue, for instance, that Ireland was ignoring its own discharges to the Irish Sea – waste from hospitals, universities and private and government laboratories. Acknowledging these discharges were small and their impact on the marine environment negligible, Britain could still point to the level of radioactivity released as far greater than any ever likely to be discharged from the Sellafield MOX plant. Furthermore, the Irish inventory included radionuclides, such as technetium-99m that decays quickly to technetium-99, one of Ireland's chief causes of complaint against Sellafield discharges at that time.

In any event, the issue of jurisdiction of the international tribunal was still open to question.

The Irish government had discarded the option of taking their case to the European Court of Justice on the grounds that it would take too long. But in their haste to get a result, the question of whether Ireland's action in going outside the EU would be in breach of its obligations under the Treaty of Rome appears to have been overlooked.

However much the Irish government sought to dress it up, there was no gainsaying the fact that by December 2003 both the OSPAR and ITLOS cases had failed. The UN case was suspended, pending clarification by the European Commission of Ireland's right to have its complaints against the UK, a fellow EU member state, adjudicated outside the institutions of the European Union. There was some consolation in that the tribunal formalised the order for continuing UK–Ireland consultation on nuclear issues. But it was increasingly obvious that the conclusion of a bilateral treaty, similar to those already in place between Britain and neighbouring European countries, was now the last remaining option.

Around Whitehall, it was difficult at first for Dublin-based embassy officials to persuade some departments that a formal bilateral agreement with Ireland was necessary or even desirable. In the years since Ireland had initiated its international court actions the mood for co-operation with the Irish authorities on nuclear issues beyond the minimum required by diplomatic courtesy had significantly soured.

Without any great fanfare or ceremony, a bilateral covering most areas of co-operation and information sharing, except transport, was signed by the British ambassador to Ireland, Stewart Eldon, on behalf of the UK government, and the Irish Minister for the Environment, Dick Roche, on 10 December 2004. The main gain for Ireland from the bilateral and the surrounding agreement was that the RPII would now have real time access to RIMNET, the early warning system for nuclear accidents in Britain. Although the bilateral could do little to diminish anti-Sellafield political rhetoric, it would at least take some of the heat out of it. The deal put 'good neighbourliness' on a formal footing, Minister for Environment, Dick Roche, wrote in the *Irish Independent*.[24]

Before long, Fergus O'Dowd was among those complaining that this new relationship on Sellafield was restricting the flow of information to the Irish public.

'There is a conspiracy of silence within the Minister's Department regarding the issue of the relationship between Britain and Ireland in respect of Sellafield and nuclear matters ... A policy of secrecy and a cloak of silence exist around [the Minister's] activities,'[25] O'Dowd said.

In January 2006, the Advocate General of the European Court of Justice delivered his opinion that Ireland had in fact broken EU law in taking its case to the UN tribunal, an opinion confirmed by the full court later in the year. Roche defended Ireland's breach of the Treaty of Rome with an attack on the European Commission.

'The Irish Government will expect the Commission to exercise its competence robustly in respect of the continued operation of the Sellafield Plant, a situation which has clearly not been the case to date,' he said.[26]

'Following consultation with the Attorney General, the case taken by Ireland against the UK under UNCLOS will not now proceed. The Commission and the UNCLOS Tribunal were informed accordingly on the 16 February last,' Dick Roche responded to a written parliamentary question on 22 February 2007.[27] The day before, unnoticed and unreported in Ireland, the ITLOS Tribunal had published its final ruling, ending the requirement for the British and Irish governments to make periodic reports on consultations.

Ireland's international court action was not the only case against Sellafield that had come to an end: on 23 November 2006 the Irish Supreme Court had finally brought down the curtains on the Louth Four case. BNFL had agreed to waive costs, believed to amount to several million euro, if the plaintiffs agreed the case should be dismissed.

* * *

Bertie Ahern achieved a notable political triumph in the re-election of his government in June 2002, the first time any Irish administration was re-elected for a consecutive term since 1969.

A decade after he had taken power in 1997, Ireland was radically changed. The main issues facing this *nouveau riche* country were how to cope with an anticipated rise in population to over five million citizens by the 2020s, the problems of coming to terms with a new multiracial society born of significant immigration, particularly from eastern Europe, economic competitiveness and related infrastructure, environment and energy supply issues.[28]

To many among the new generation of Irish voters facing the 2007 election and the decision as to whether or not to accord Bertie Ahern a third term as Ireland's leader, the Sellafield issue was more a childhood memory from their schooldays when they had been compelled to write essays endorsing the national anti-nuclear policy than an issue of immediate pressing concern. Global warming, quality of life issues and the strategic direction of a country, prosperous, proud and at peace with itself as well as Northern Ireland, were of greater import in guiding their political choices.

Ahern's government had fared well on economic policy and the resolution of the Northern Ireland question; but the environment had been almost brutally neglected, of which the long-fingering of solutions to Ireland's own nuclear waste provided just another example. The focus on Sellafield, and the scaremongering that accompanied it, may have induced a strong anti-nuclear bias among the Irish population. At the same time, it made Ireland's own specific nuclear legacy more difficult to manage.

By 2006, there were 1,400 holders of sealed sources in Ireland, licensed by the RPII. Of these, 943 were in use in the irradiation cells of three sterilisation plants and the remainder distributed throughout Irish industry. Ireland had a total of 6,159 disused radiation sources. The fuel slugs from the UCC reactor accounted for about 1,400 of these. The remainder were stored at about seventy different locations throughout the country.[29]

To combat the risk of such materials coming into the possession of terrorists and used in production of a so-called 'dirty bomb', the European Union in 2003 adopted the HASS Directive[30] which, among other things, required member states to make suitable arrangements for the storage, monitoring and safeguarding of 'orphan sources' within their respective territories. The RPII estimated there were about 600 such 'orphan sources' in the country, a small number of which would be classified as highly radioactive, energy official Renee Dempsey had informed the Oireacthas Committee on the Environment in January 2004.[31]

Ireland would have to ratify the HASS Directive by the end of 2005 and the RPII had for some years been examining three or four sites for a national repository to house this waste.

'An option increasingly being suggested by expert bodies, such as the International Atomic Energy Agency, is that the sources be placed in boreholes in the ground, which are then backfilled with concrete. The boreholes would be typically 15 centimetres in diameter and 100–300 metres deep,' Ms Dempsey said, adding that 'most sealed sources will decay to background levels in a few hundred years'.

'I understand there will be public concern about bringing together all the sources,' the RPII CEO Dr Ann McGarry acknowledged, 'but in terms of the radiation hazard that presents, it is our view that the present dispersed locations present a greater hazard.'[32]

'It will be a difficult nut to crack but it is an obligation,' Renee Dempsey concluded.[33]

As it turned out the obligation took second place to the difficulties with the nut: the Minister for Environment, Dick Roche, dealt with the obligation by means of a Statutory Instrument in late 2005, a legislative mechanism that was laid before parliament but not debated.

S.I. 875 of 2005 transposed the HASS Directive into Irish law. It designated the RPII as the agency responsible for implementing the directive but sidestepped the question of any national repository. Instead, the RPII would deal with orphan sources on a 'case by case' basis with the Department of Environment, Heritage and Natural Resources.

'The practical management of orphan sources including their storage, particularly where no legal entity can be identified to take custody of them, remains a serious issue to be addressed and continues to be a weakness in the overall regulatory infrastructure in Ireland,' the RPII noted in its Annual Report for 2005.[34]

* * *

The NDA had estimated the clean-up costs for the Sellafield site at around £36bn in 2005. Thereafter, the costs increased. Removing the sludge tanks, the 1940s-built silos for nuclear waste; reducing the inventory of 400 million curies of highly active liquid waste and stabilising it in glass; tackling the radioactive sewer represented by the

B30 pond; securing and demolishing old buildings that provided, among other things, nesting sites for the most radioactive feral pigeons on the planet; decommissioning the Calder Hall and WAGR reactors; finding a solution to the 100 tonne plutonium mountain on the site – all these and more comprised the problems and priorities for Sellafield in early 2007.

There were also, of course, the old Windscale piles, still waiting decommissioning on the fiftieth anniversary of the 1957 fire.

The Windscale fire has retained its fascination for researchers throughout the fifty years since the accident. Among the most recent studies on the atmospheric distribution of the plume, a 2007 paper indicated that the plume that had covered all of England and Wales by 14 October 1957 and had drifted eastwards over Europe had, at most, brushed the east coast of Ireland.[35] In March 2005, a UCD team of scientists published a research paper showing that no fallout from the Windscale fire had, in all probability, ever reached Ireland.[36] The scientists examined sediments from Ballywillan Lake in Co. Down, using iodine 129 as a tracer for possible earlier fallout of iodine 131 from Windscale. 'The study yields no evidence of any enhancement in radioisotope concentrations, over and above global fallout, in strata dated to 1957, and we conclude that contamination from the Windscale fire had negligible impact on the northeastern region of Ireland,' they stated.[37]

Among the Sellafield shop stewards, to those who became aware of it, such news came as a relief. Their operations were now proven not to have caused any damage to Ireland, they believed, a country to which several of them felt a close and familial connection. They recalled the story of one of their shop stewards, a keen angler who regularly took his holidays in Ireland. By instinct more than design, he had adopted the habit of refusing to disclose his true occupation, mentioning only that he lived in Cheshire and worked in some nameless factory there. On a trip to the lakes of Killarney in the 1990s when the fish had resolutely refused to bite, his boatman had apologised for a wasted day. 'God's truth,' the boatman said, searching for an explanation, 'it's probably Sellafield's to blame.'

By 2007, the NCNI, now renamed Nuclear 21, had left old battles far behind and were busy campaigning for one or even two nuclear power stations at Sellafield. Burning MOX, the new stations could, over their operating lifetime, dispose of the circa hundred tonnes of plutonium that had built up at the site over fifty years, the unions claimed. Moreover, new build at Sellafield would preserve the skills base of Britain's north-west nuclear workforce and the jobs of their community.

* * *

As of January 2007, approximately four thousand tonnes of graphite from the Windscale piles awaited removal and decommissioning. In the fire-damaged Pile No. 1, the fifteen tonnes of fuel damaged in the fire and always at risk of combustion remained undisturbed.

Decommissioning, the NDA said, would take place in two stages: first, dealing with the remaining inventory of radioactively contaminated materials by 2013; then, more than sixty years since Windscale had produced the first plutonium for Britain's nuclear bomb test at Monte Bello, removal of the reactor cores and of the last traces of the Windscale piles would begin. The expected date of completion: 2063.

22

Conclusions

THE LEGACY OF DISCOVERY

The pioneering scientists of the nuclear age, theoretical and experimental physicists, chemists and mathematicians, from the end of the nineteenth century to the 1930s, enjoyed a brief era of complete intellectual freedom. Their insights into the nature of the physical world at its smallest level owed much to their individual and collective obsessions with expanding knowledge for the benefit of all humanity.

Our revolution in information technology is part of their legacy. As is much of modern medicine: radiological and diagnostic therapies used in the treatment of various cancers, heart disease and many other ailments save millions of lives each year and give hope of extended life to millions more around the world. Without nuclear technology the safe manufacture of many products for human use and consumption could not be assured. Throughout Europe, the US and Asia, nuclear energy makes a stable and often vital contribution to electricity supply in many countries. But ten years before Rutherford's announcement in Manchester in 1911 of his discovery of the atomic nucleus, scientists were also aware of the destructive potential of atomic power, the mathematical basis for which was embodied in Einstein's 1905 formula, $e=mc^2$.

THE BOMB MAKERS

Retrospective condemnation of the nuclear scientists for their role in development of the atomic bomb is not historically justified. The rise of fascism in Europe in the 1930s and, in particular, the spread of Hitler's hegemonic grip on the continent forced many scientists and their families, in fear of their lives, to flee their homes and countries. As a refugee group they were unusually fortunate: these members of the 'independent republic of science', so to speak, benefited from a network of contacts that had developed internationally over the previous three decades. There was refuge and relative safety to be found in Sweden and for a time in Denmark and France, and latterly in the universities of Britain and on the campuses of prestigious American colleges. When war came, many among them, as we have seen, relished the prospect of contributing their knowledge and expertise to the defeat of Hitler and fascism in Europe.

The bombs that obliterated the two Japanese cities of Hiroshima and Nagasaki in August 1945 could not have happened without the scientists; but the bombing didn't happen because of them. If the first casualty of war is truth, the restriction of individual rights and freedoms runs a close second. By their very nature wars escalate beyond any original grand strategic objectives in a cycle of increasing brutality. That innate moral compass that guides our individual thoughts and actions towards others collectively jerks out of control. As the Second World War demonstrated on a global scale, in the end it is not just the rights of the individual that are cast by the wayside; genocide and mass murder on an industrial scale can become institutionalised values even of the most ostensibly civilised peoples, the deployment of weapons of mass destruction an acceptable tactic to defeat an enemy already dehumanised in popular opinion.

As recounted in the early chapters of this book, only one country, the United States, had the resources in wartime to carry an atomic bomb programme to fruition, and once the bombs were made, determination to demonstrate their effectiveness. Appalling and all as it may seem in the light of our modern sensibilities, Hiroshima and Nagasaki represented little more than collateral damage in the pursuit of longer-term strategic political interests.

The same mix of social attitudes and political influences determined the mindset of the Manhattan project scientists as anyone else in society. Among many of the refugee scientists their political inclinations veered towards a centralised global system of government, based on socialist principles. Leo Szilard appeared as influenced by new world order ideas contained in the writings of H.G. Wells as he was by Wells' theoretical musings on the possibility of constructing nuclear weapons; Edward Teller, as late as 1948, was writing of the need for the scientists in the post-war era to work for World Government.

Szilard spent the remaining years of his life striving to squeeze the nuclear genie back into its bottle. Teller went on to become the father of the hydrogen bomb and originator of the 'Star Wars' concept.

Ultimately, he is credited with having provided inspiration for Stanley Kubrick's 1964 film portrayal of the mad scientist in *Dr. Strangelove*.

The Manhattan scientists were a product of their times and generation. Their choices and actions reflected their individual personality preferences and political beliefs at any given moment, whether during the Second World War or in the Cold War that followed it.

WALTON

Speaking of his own father's views on the development of atomic weapons, Professor Philip Walton noted: 'He wasn't an absolute pacifist in that he felt if the evil regime of Hitler was going to get the bomb, he wasn't absolutely anti (the Manhattan project). In retrospect, he was glad that he hadn't been personally involved, but at the time, it would have been different.'

The decision to join the project to build an atomic bomb was never in the gift of the Irish physicist. Had he been in a position to respond positively to Oliphant and C.P. Snow's offers, Ernest Walton's subsequent career might have taken a different course. As it was, Walton the experimental physicist became Walton the revered teacher, as well as devoted husband and father to a family of four children all of whom went on to enjoy successful lives and careers.

Throughout his tenure as a Professor of Physics in Trinity College, Walton sought no public recognition or advancement on foot of his distinction as Ireland's only Nobel laureate in science. 'He wouldn't really have cared much about public recognition anyway,' Philip said.

Walton's life history and experiences illuminate many aspects of the development of Ireland, and its nuanced relations with Britain and the rest of the world, from the early years of independence. Remembered by all who knew him as a man of absolute integrity, with a deep and sincere commitment to the welfare of Ireland and to Trinity's progress and development, Walton, after his retirement in 1974, maintained his links with the college. Shortly before his death at a nursing home in Belfast in 1995 he presented his Nobel medal and citation to Trinity.

BRITAIN'S ATOMIC BOMB PROJECT

The British scientists at Chalk River persuaded by Walton's co-laureate, John Cockcroft, to return home to work on Britain's post-war independent nuclear deterrent programme were less concerned about salving their consciences for any role they played in producing the world's first atomic weapons than about saving their country from the newly perceived threat of Soviet aggression.

The architects of Britain's early nuclear military programme, Cockcroft, Chadwick, Penney and Hinton, were driven by the idea that Britain must secure a lead in post-war nuclear technology, civil and military. They perceived this as absolutely necessary, both in the long-term interests of the preservation of democracy and freedom as well as the future prosperity of their country.

The coalition of 'Cold Warriors' extended to the West Cumbrian workforce at Windscale producing the base materials that were turned into atomic bombs. All active trade unionists, part of a movement with a communist leadership, they were none the less aware, as Jim Innes put it, 'that there was a subtle link between what they were doing and the need, perhaps one day, to confront the Red Army'.

Britain's post-war political leaders, making decisions about bomb-building without reference to the wider cabinet, party, parliament or people, had Britain's national defence as their first priority. Decisions were complicated by an obsession with Britain's post-war international status as a world power and securing the nuclear dimension of its 'special relationship' with the United States as a key objective of foreign policy.

Echoes of the workings of that phase of the special relationship arguably reverberate in our own time. Not least, perhaps, in the extent to which British post-war governments set aside taking leadership in Europe on the development of nuclear technology to pursue the special relationship with the United States. Instead of placing Britain at the heart of Europe, they set their store by the United States, a relationship in which they were all too often humiliatingly relegated to the status of 'poor relation' and treated accordingly.

THE POLITICS OF SELF-INTEREST

The pre-requisite of any nuclear weapons programme is secrecy. The commercial possibilities of nuclear technology, such as large-scale development of civil nuclear energy, were expected to confer huge advantage on any country exploiting them. Research and development of the applications of civil nuclear technology were thus almost as closely guarded as weapons development – most marked in the initial

US preoccupation with maintaining a monopoly and its reluctance to agree any technology transfer arrangements with its European allies, including Britain. 'Atoms for Peace' ensured American control of the development of commercial nuclear technology amongst US-allied countries. As a Cold War stratagem, it separated the US-allied sheep from the Soviet goats. Commercially, it gave American civil reactor designs a competitive edge in future markets in the 1960s and 1970s.

In the immediate post-war era, as now, countries pursued a nuclear policy based on calculations of their own self-interest. The development of NATO obviated the need for independent atomic weapons development in most of Western Europe. In Britain, as we have seen, Calder Hall was initially envisaged as a purely civil nuclear project. The requirements of the military programme would always take precedence over the development of the civil uses of nuclear technology.

Britain appeared to have little difficulty in supporting the nuclear ambitions of its 'White Dominions' among the Commonwealth countries and others such as Israel, whether civil or military, whilst discriminating against the rest. With hindsight, it seems ironic, and hypocritical, that Ireland's wartime neutrality and De Valera's adherence to that policy as in the best interests of a small defenceless country, as he himself characterised it, should have drawn so much opprobrium from Churchill throughout the course of the Second World War and afterwards from post-war US administrations.

IRELAND'S NUCLEAR HISTORY

The Sinn Féin revolution failed to deliver the self-sufficient, de-anglicised, Gaelic Ireland of which its ideologues, whether in Cumann na nGaedheal or Fianna Fáil or scattered elsewhere along Ireland's political spectrum, had dreamed. Instead it gave rise to a moribund establishment of rigid, uncompromising and competing elites whose resistance to any social or economic reforms that might in any way dilute their privileged position in society brought the state almost to the brink of collapse by the mid-1950s.

That Ireland, and indeed Northern Ireland, as illustrated by Glentoran's escapade in the 1950s to secure location of a UKAEA nuclear installation of any type for the North, was keen to avail of any advantages that might accrue from the use of new prestigious technologies is hardly surprising.

Ireland couldn't afford to take up the US 'Atoms for Peace' offer to purchase a research reactor in the 1950s. Ernest Walton recognised this and strongly advised that acquiring a reactor should remain down the order of Ireland's development priorities. This did not imply any reservation on his part to the use of nuclear power.

'Our first nuclear power station will be required about ten years hence and we are unlikely to need a second for a long time afterwards,' he had written to his colleagues on the Atomic Energy Committee.

'When the time for placing the contract for the first one is approaching, we should entice back a few of our nuclear engineers who have been employed (abroad) in the nuclear power industry.'

When that time did come, with the first oil shock of the 1970s, the decision to go nuclear was forced on an Ireland already 75 per cent dependent on oil. The full story of the ESB proposal to build a nuclear power station at Carnsore in Co. Wexford is instructive on what can happen a small country, always at the end of everyone else's pipeline, in a global energy crisis. The Carnsore story further exposes the political opportunism of those who have retrospectively demonised the then Minister, Des O'Malley, and government as 'nuclear zealots' while casting their own environmental credentials in the hazy glow of anti-nuclear populism.

As in the late 1950s, Ireland in the late 1970s could not afford a nuclear power station as it pitched forward into a decade of financial crisis and negative economic growth. The Three Mile Island accident, not popular protest, finished off the Carnsore project. A nuclear power plant, built at a capital cost of £500m, that might, within a matter of weeks, meet a similar fate to TMI-2 leaving a clean-up bill of twice the original cost, was not an option for any Irish government of the time.

SHUT SELLAFIELD POLICY

Ireland's campaign to close down Sellafield owes its origins, as much to the post-colonial relationship between the two countries, including Ireland's persisting economic dependence on trade with Britain, its inability to break the ties of language, politics and culture that so frustratingly bound it to its larger neighbour, and the unresolved issue of the partition of Ireland as to any anti-nuclear philosophy. Charles J. Haughey unashamedly sought to use Sellafield as an issue to drive a wedge between the Taoiseach, Garret FitzGerald and the British Prime Minister, Margaret Thatcher, as part of his opposition to the 1985

Anglo-Irish Agreement. As an exercise in political populism, FitzGerald's government could be lampooned as pro-British, as fearful of confronting Thatcher on an issue that many people in Ireland were genuinely convinced was damaging the health of Irish people and a major threat to its environment. Thereafter Irish politicians of all parties were quick to realise that Sellafield was a 'safe' political issue that would not cause any disruption to the conduct of business in other more vital areas of Anglo-Irish relations, such as Northern Ireland.

FALLOUT FROM WINDSCALE '57

It has now been shown fairly conclusively that the Windscale fire had no radiological consequences on Ireland, but came to have a marked psychological impact on public attitudes to nuclear power. Especially so following the publication of Dr Patricia Sheehan's discovery of the Down's syndrome cluster in Dundalk and the suggestion in her 1983 BMJ paper that the cluster might be linked to fallout from the Windscale fire.

Momentum for Ireland's political campaign against Sellafield derived from widespread public fears for the health of Irish citizens and the implications for Ireland, its environment and the Irish Sea, of the expansion of Sellafield's commercial operations in the 1980s and 1990s. The accident at Three Mile Island and the 1986 Chernobyl disaster compounded public opposition in Ireland to nuclear power.

Early Irish expressions of concern over issues of health and safety relating to the operation of Sellafield were the actions of a responsible government and entirely justified in the context of events at the site and fears about the effects of routine radioactive discharges or a major accident on Ireland. But many of the leading figures of Irish politics, from the foundation of the state to the 1980s, would have been rightly perplexed by the refusal among the current political generation to make any distinction between the civil and military uses of nuclear technology. In fact, it was not until the mid-1990s that this distinction finally disappeared and with it an acknowledgement of the many beneficial uses of nuclear technology, in industry, science or medicine.

BRITISH RESPONSE TO IRISH CONCERNS

The means selected for expressing Irish concerns about the Sellafield complex over time became ever more strident and uncompromising.

Britain has a long tradition of resistance and resentment against any external efforts to interfere with its conduct of nuclear policy. British officials dealing with Ireland were inclined to regard vociferous protests by Irish politicians as more reflecting the cut and thrust of local inter-party rivalries, designed for domestic political and media consumption, or as a safety valve for historical anti-British resentment, than as based on any genuine source of grievance.

To some extent, that perception was probably justified. But in some quarters of the British establishment Ireland's protests were routinely greeted by British officials with a scarcely veiled mixture of contempt and disdain. Up to comparatively recent times, there were many in Whitehall and among the British political establishment who appeared to have real difficulty – much as did their predecessors throughout the early years of Irish independence and the Second World War – in reconciling themselves to Ireland's status as a sovereign independent country, a close neighbouring state with a well-founded interest in developments at Sellafield and the British nuclear industry generally. Responding to Ireland's representations on nuclear matters with dignity and respect rather than, as was all too often the case, with indifference and denial might have proved a more fruitful policy.

Following the Thorp campaign, BNFL woke up to the reality that Irish opposition to Thorp and Sellafield's commercial operations might damage its prospects and reputation as an internationally trading nuclear services company. As a state-owned enterprise, BNFL was subject to diktats from the Foreign Office, its parent department and regulators. When it came to BNFL's policy towards Ireland, the reactions at official level were often counterproductive and seemingly based on ill-informed and biased interpretations of Irish concerns. British ministers and parliamentary committees showed insensitivity, often appearing oblivious and entirely indifferent to the effect of their decisions on public opinion in Ireland.

The most outrageous and ill-founded allegations were, as Brian Wilson has pointed out, also allowed to pass unchallenged by successive British governments. But few, if any, attempts were ever made to address Irish concerns through any process of constructive engagement, until it was almost too late. The UK government brought Ireland's international court actions of the early years of this century down on its own head.

DEVELOPMENT OF BRITAIN'S NUCLEAR SERVICES INDUSTRY

Windscale was the hub of Britain's military nuclear programme. In the second phase of its existence, it became a dual purpose civil/military operation.

As recounted in this book, plutonium warheads were reprocessed at Sellafield to remove americium contamination and the Calder Hall reactors continued their function of producing military-grade plutonium and other weapons-related materials for several decades. Simultaneously, Sellafield was expanding its civil nuclear services, especially reprocessing and waste management.

By the mid-1990s, Sellafield, as confirmed by then Foreign Secretary, Douglas Hurd, at the 1995 UN NPT negotiations, was no longer in the business of producing fissile materials for nuclear weapons. With Thorp up and running, BNFL was poised to enter a new commercial phase, providing lucrative reprocessing services to a range of international customers, completing the nuclear fuel cycle by turning the recovered materials into fresh MOX fuel.

A lot of hope in the post war years was invested in civil nuclear technology as a future energy source – supplies of cheap Middle East oil had not come on stream to any great extent at that time and Britain's own reserves of North Sea oil were not yet discovered. Ordinary citizens were also considerably more deferential both to their political leaders and 'expert opinion'. That deference began to dissipate in the 1960s and early 1970s and public confidence in 'expert opinion' has become progressively even more depleted in more recent times.

The emergence of environmental pressure groups, such as Greenpeace and Friends of the Earth, who campaigned against the development of nuclear power, combined with accidents such as Three Mile Island and Chernobyl, and the apparent inability of the British nuclear industry to get anything right, all contributed to a general loss of public confidence in nuclear energy.

'The problem with the British Nuclear industry for a long time has been the anti-nuclear weapons debate,' Brian Wilson suggested.

'In the early days a distinction was made on the political left between the military and the civil use of nuclear power, but at some point that argument was conceded and they got lumped in together. So being anti-nuclear, quite irrationally, covered civil nuclear power as well as military, and whatever the chronology of it, once that ball started rolling there were plenty of episodes to keep it moving along.'

This is not the whole story. The development of Britain's nuclear industry, unlike France for example, was bedevilled by a lack of political consensus. Politicians dithered in the 1960s, sometimes for ideological reasons, about the best choice of reactor, and according to most experts picked 'the wrong one' in their preference for the home-grown over-designed and over-engineered AGRs. In France, a clear political decision to run with US designs was adopted.

Arguably, responsibility for the British nuclear industry was spread across too many departments and too much power over the direction of policy was concentrated among the experts in the UKAEA and later in BNFL. When the Department of Energy ceased to exist in its own right, no minister ever stayed, or was left, long enough in the energy portfolio to introduce policy coherence. The privatisation model adopted in the Thatcher era, the 'dash for gas' of the 1990s, the pool model for wholesale electricity pricing and political uncertainty made further investment in nuclear uneconomic, and British Energy was bankrupted.

Greenpeace and Friends of the Earth capitalised on the divide in British political opinion over nuclear policy and placed BNFL and Sellafield at the centre of their international anti-nuclear campaigns. In the war of attrition between BNFL and the international environmental lobby for the best part of thirty years, the result was an uncomfortable draw for both sides. Public opinion in Britain, and elsewhere with the exception of Ireland, never wholly swung behind the Greenpeace campaigns. BNFL fought off Greenpeace and Friends of the Earth in court actions in the UK and internationally. The primary objective of the environmental organisations, to shut down Sellafield, was not achieved.

In highlighting the deficiencies in the plans and programmes of nuclear management and the build-up of wastes and hazards at Sellafield, the environmental groups, however, performed a valuable public service. Lack of investment in containment and waste management systems at Windscale in the early years and the political failure by successive governments to either acknowledge or prioritise the waste issue contributed to Sellafield's later difficulties with leaks and discharges and failing installations.

In the early years of Britain's nuclear programme public acceptability was a given. The Windscale piles had been conceived as a national necessity. Public opinion was not a factor, either in the purpose of the Windscale site or its industrial operation. Old ingrained habits died hard in the nuclear industry. By the time public acceptability was recognised as crucial to the industry's survival, a legacy of attitudes, in which an isolationist siege mentality became increasingly paramount, had grown up to match the legacy of historic wastes at Sellafield.

IRELAND'S CAMPAIGNS

In a microcosm of the broader international struggle between environmentalist NGOs, local campaign groups and the nuclear industry worldwide, Ireland and Britain played out their political battle over Sellafield. Yet, Ireland's political and celebrity campaigns to close Sellafield achieved nothing of substance. David met Goliath, but Goliath strolled on regardless.

Given the nature of the hazards on the Sellafield site, closure as demanded by Irish politicians was never an option. Without Sellafield, the magnox stations, vital to Britain's electricity supply, could not operate, since reprocessing spent magnox fuel was the best economic and environmental strategy available at the time. Sellafield also provided the primary waste management and storage facilities for the whole of Britain's nuclear industry.

Premature closure of the site, in any case, which would leave the plutonium mountain to rot, the silos and tank farms and the remainder of the historic legacy on the site untreated and unmanaged and necessitate the repatriation to European countries and Japan of several thousand tonnes of unreprocessed nuclear fuels, was hardly in Ireland's interest, even if it was possible.

Irish politicians and media justified their demand to shut Sellafield on the basis that Ireland derived no benefit, and all the disadvantages, from Britain's nuclear industry. This too was false, since it ignored the economic interdependence of the two countries.

It also ignored Ireland's sourcing of nuclear materials in Britain for use in medicine, industry and research and, in many cases, their ultimate disposal route, back to Drigg. As Fine Gael's Bernard Allen observed in the mid-1990s: 'I went to Sellafield in 1983 and asked pertinent questions in relation to its operation. I was politely taken down the road to a dump which was called a treatment plant and shown low and medium radioactive waste exported from Ireland. It was made quite clear that we were dependent on them to dispose of some of our nuclear waste and the position has not changed ... we are totally compromised in our stance with the British Government on this issue.'

The Kyoto Treaty obligations to cut greenhouse gas emissions exposed another strand of interdependence between the two countries in which the nuclear issue is a factor. Without renewing its nuclear capacity, Britain may experience difficulty in meeting future climate change targets set by the EU for the post-Kyoto Treaty era. The best hope for small EU states like Ireland to secure reasonable emission targets that will not damage our future development and prosperity relies on the larger countries like Britain being willing and able to take up the slack.

Ireland's political campaign from the mid 1990s increasingly focused on forcing the closure of Sellafield's international commercial operations – Thorp reprocessing, nuclear materials transport, conversion of reprocessed plutonium to MOX fuel and, as we have seen, a determined effort to stop construction of any national waste storage facility. The policy of successive Irish governments was driven more by anti-nuclear ideology than expert opinion or advice as to the real threat, if any, posed to Ireland by the activity being proposed.

IMPACT OF SELLAFIELD ISSUE ON IRELAND

Living next door to Sellafield generates considerable unease, and very real fears, among Irish people, especially those living along the east coast and in Co. Louth. Ireland's nuclear watchdog, the RPII, has consistently sought to provide a measured and factual picture of the hazards posed by Sellafield to Ireland and its environment. If anything, their task is becoming more difficult, as security restrictions in Britain and the dissolution of the informal relationships which RPII officials enjoyed with BNFL limit their capacity to make a comprehensive analysis of safety issues at Sellafield.

Too often in this story they have been a lone voice, lost among a media and political chain reaction of outrage and hysteria that, in the end, served little purpose except the generation of unreasonable fear.

Ireland's uncritical adoption of the Greenpeace agenda as national policy on nuclear issues and Sellafield, endorsed by the media, represents a major triumph for the environmental organisation and John Bowler's lobbying skills. What has to be questioned is whether that agenda, while it may have been entirely appropriate to an international NGO, wholly served the best interests of a sovereign country. Agreements between Britain and Ireland on nuclear information sharing and consultation would most likely have been concluded much earlier if a different approach had been followed.

Demonstrably, BNFL's fall from grace came about as a result of the actions of the company's own management and UK government ministers, either inadvertently or deliberately, that had nothing to do with Irish opposition to Sellafield. For all its defects and shortcomings, and its regular demonisation by Irish politicians and in the Irish media, BNFL as the sole operator of the Sellafield site has belatedly come to be viewed as a preferable guardian of safety at Sellafield than a fragmented division of responsibilities for the site among a number of private contractors. As an *Irish Times* editorial on 31 March 2006 noted: 'That the operation of Thorp and the clean-up of Sellafield was entrusted to a State-owned body, albeit one with a dodgy record, offered a small measure of comfort. For it to be handed over to profit-maximising private interests will add to the unease.'

Sellafield served as a distraction for politicians from having to deal with Ireland's own environment problems. Designating Sellafield as the greatest environmental threat to Ireland somehow made national environmental issues appear less threatening or urgent. It also made them more difficult to deal with. At the start of 2007, the Irish government was the subject of some thirty infringement proceedings by the EU for failure to observe or properly implement European environment laws.

As the bizarre mechanism chosen for implementation of the EU HASS directive shows, anti-Sellafield posturing has limited Ireland's ability, to the point of political impossibility, to face up to its responsibilities to deal with our own relatively insignificant – at least on the Sellafield scale of things – nuclear waste problem.

Ireland's international court cases, too, ended in failure and in retrospect appear reckless and naïve. The rationale for the change in twenty years of advice from successive Attorneys General that such a case should only be brought if it stood a reasonable chance of success has never been satisfactorily explained.

The possibility of one further international court case remains open – perhaps relating to Britain's contamination of the shared resource of the Irish Sea over a prolonged period – but any Irish government embarking on such a course of action would need to feel assured of a successful outcome.

CHANGING ATTITUDES

Based on studies to date, especially more recently published work, there is no scientific evidence to support allegations of any detrimental effects of Sellafield operations on public health or the environment in Ireland. In particular, the Windscale fire of 1957 has been absolved of any blame for incidence of Down's syndrome or other health effects in Ireland. Among the political classes, the Sellafield issue can no longer serve as a fig-leaf to cover up the failure to address pressing local environmental issues. It remains to be seen if Irish concerns about Sellafield and its operations and any threats, real or imagined, that it poses to Ireland can be more easily accommodated within the now more cordial framework of Anglo-Irish relations than in the past.

In Ireland, as elsewhere, climate change is also forcing a re-evaluation of old prejudices against technologies, such as nuclear, that offer immediate solutions to the problems of large scale emissions of greenhouse gases. There is a recognition that there is no wholly safe way of mass producing energy and that all systems, including renewables, carry environmental and economic costs. As the debate continues, the story of Ireland's nuclear dispute with Britain is instructive: in the long run, fundamentalist posturing on either side of the argument did no one any good.

Notes

NB Where a short version of a book title is given, the full publication details can be found in the Select Bibliography.

1. NEW BEGINNINGS

1. 'Obsessive Historian: Eamon De Valera and the Policing of his Reputation', Patrick Murray, Royal Irish Academy, Dec. 2001.
2. Letter from Frank Hackett to Desmond FitzGerald, quoted in Tom Garvin, *Preventing the Future. Why was Ireland so Poor for so Long?* (Gill & Macmillan, 2004), page 39.
3. L.M. Cullen, *An Economic History of Ireland since 1660*, (London: B.T. Batsford, 1972) page 171.
4. Dáil Debates 1927–1932, Economic Debates.
5. Garvin, page 2.
6. See Diarmaid Ferriter, *The Transformation of Ireland 1900–2000* (London: Profile Books, 2004), pages 44–7.
7. Ibid.
8. See *Dáil Reports*, Vol. 41, 21 Apr. 1932, Col. 328. Lemass to Mc Gilligan: 'It has been estimated that in past years the average number of migratory labourers going to Great Britain to insurable [328] employment was about 14,300 and to non-insurable employment of various kinds about 10,725, but the information necessary for more precise particulars at the present time would take a considerable time to obtain.'
9. Walton later recalled having lived in ten different locations in Ireland throughout his youth, six of which were in Northern Ireland.
10. The Solemn League and Covenant read as follows: 'Being convinced in our consciences that Home Rule would be disastrous to the material well-being of Ulster as well as the whole of Ireland, subversive of our civil and religious freedom, destructive of our citizenship and perilous to the unity of the Empire, we, whose names are under-written, men of Ulster, loyal subjects of His Gracious Majesty King George V., humbly relying on the God whom our fathers in days of stress and trial confidently trusted, do hereby pledge ourselves in solemn Covenant throughout this our time of threatened calamity to stand by one another in defending for ourselves and our children our cherished possession of equal citizenship in the United Kingdom and in using all means which may be found necessary to defeat the present conspiracy to set up a Home Rule Parliament in Ireland. And in the event of such a Parliament being forced upon us we further solemnly and mutually pledge ourselves to refuse to recognise its authority. In sure confidence that God will defend the right we hereto subscribe our names. And further, we individually declare that we

have not already signed this Covenant. The above was signed by me at ... "Ulster Day". Saturday. 28th September. 1912. God save the King.'
11. The college provided free board to the children of Methodist ministers to ensure the continuity of their education, which might otherwise have been jeopardised by the three year rule.
12. Trinity's eminence as a seat of learning in mathematics and science dated back to the nineteenth century and Sir William Rowan Hamilton, Ireland Astronomer Royal and inventor of the Hamiltonian method in mathematics.
13. Richard Rhodes, *The Making of the Atomic Bomb*, pages 37–8.
14. In 1901, Roentgen was awarded the first ever Nobel Prize in Physics for his discovery. His X-rays were later correctly identified as of a similar type to gamma radiation.
15. See Rhodes, *The Making of the Atomic Bomb*, for comprehensive account of early history of theoretical physics.
16. The Russian chemist Dmitri Ivanovich Mendeleev first published the Periodic Table in 1869.
17. The element thorium had been identified in 1829 by a Swedish chemist, who named it after the Nordic god Thor.
18. In fifty years of radium extraction from uranium ore, only five kilogrammes were produced throughout the world. Radium, with its half-life of 1,620 years, was recognised as highly toxic with less than five microgrammes constituting a serious risk to health. See Margaret Gowing, *Independence and Deterrence* Vol. 2 (London: Macmillan Press, 1972), pages 111 and 314.
19. Gamma radiation was later identified by the French physicist Paul Villard, working on cathode-tube rays in Paris in 1900. Rutherford later named these more penetrating rays 'gamma rays' and in 1914 showed they were similar to EM or X-rays though of a shorter wavelength.
20. Frederick Soddy was awarded the 1920 Nobel Prize in Chemistry for his work on radioisotopes and the periodic table.
21. Quoted in Rhodes, *The Making of the Atomic Bomb*, page 44.
22. Eugene Geiger, later inventor of the Geiger counter, and Ernest Marsden conducted experiments on the scattering of alpha particles under Rutherford's direction at Manchester in 1908, which led to Rutherford's first intuitive conclusions concerning the structure of the atom.
23. Rhodes, *The Making of the Atomic Bomb*, page 50.
24. Polanyi was Professor of Physical Chemistry at Manchester University (1933–48) and Professor of Social Science (1948–58).
25. Quoted in Rhodes, *The Making of the Atomic Bomb*, page 34.
26. D.J. Murnaghan, 'History of Radium Therapy in Ireland', *Journal of the Irish Colleges of Physicians and Surgeons*, Vol. 17, No. 4, Oct. 1988.
27. Preston, Diana, *Before the Fall Out* (New York: Doubleday, 2005), pages 26–7.
28. In all, 1,413 people were killed in air raids on Britain during the Great War, 670 of them in London.
29. Chlorine and 'mustard' gas, dichlorethyl sulphide, was first used on the western front by the Germans in July 1917. One of its effects was to cause temporary blindness.
30. Ironically, Rutherford was notorious for his snobbery about all other branches of science than physics, which he held was true science, the rest mere 'stamp-collecting'.

31. Quoted in V.J. McBrierty, *Ernest Thomas Sinton Walton 1903–1995: The Irish Scientist (Dublin: Trinity College Dublin Press)*, page 27.
32. In 1899, Joly had used the rate of sodium deposits in the sea to estimate the age of the earth, which he calculated at about 90 million years old. Later radio metric calculations suggested, accurately, that the Earth is approximately 4.5 billion years old.

2. SPLITTING THE ATOM

1. The Australian physicist Mark Oliphant, who also arrived in Cambridge in 1927, described Rutherford's pipe as belching 'smoke and ash like a volcano'.
2. Rhodes, *The Making of the Atomic Bomb*, page 141.
3. Schrödinger's theory ultimately provided the foundation for the development of semi-conductors and laid the basis for the modern age of technology.
4. Rhodes, *The Making of the Atomic Bomb*, page 133.
5. One of Walton's favourite anecdotes concerned the difficulty the Chief Technician experienced securing permission from Rutherford to purchase a bucket. When he was told the bucket had a hole in it, Rutherford boomed: 'Well, can't you mend it?' The Director only relented when it was pointed out the bucket had been mended three times already.
6. Geiger later developed his counting machine into the Geiger counter for measuring radioactivity, which is still used today.
7. Cathcart, Brian, *The Fly in the Cathedral*, 2nd edn (London: Penguin, 2005), page 99.
8. Metropolitan Vickers, one of Britain's foremost industrial engineering companies by the 1920s, originated as British Westinghouse, founded by the American George Westinghouse, in 1899.
9. Quoted in Cathcart, *The Fly in the Cathedral*, page 63.
10. A full account of Chadwick's experiments is given in Rhodes, *The Making of the Atomic Bomb*, pages 163–5.
11. Brian Cathcart, *The Fly in the Cathedral*, page 162. Walton's £250 DSIR grant was supplemented by a £200 memorial scholarship which was within Rutherford's own gift.
12. Cathcart, *The Fly in the Cathedral*, page 170.
13. Quoted in McBrierty, *Ernest Thomas Sinton Walton*, pages 28–9.
14. Ibid.
15. That same month, an American team observing cosmic radiation and using Gamow's wave theory had identified the positron, the positively charged companion particle to the electron in the atom's outer shell.
16. Quoted in Cathcart, *The Fly in the Cathedral*, page 250.
17. Bowman, 'Eamon De Valera: Seven Lives', in *Eamon De Valera and his Times*, edited by J.P. O'Carroll and John A. Murphy (Cork University Press, 1986).
18. *Morning Post*, editorial of 22 February 1932. Quoted in Bowman, ibid. page 182. The *Morning Post* merged with the *Daily Telegraph* in 1937.
19. Ibid.
20. Maryann Gialanella Valiulis, 'The View of the Opposition: Eamon De Valera and the Civil War', in O'Carroll and Murphy, *De Valera and his Times*, page 92.

21. Many of the Protestant ascendancy had Catholic origins. The Catholic share of landed estates had fallen from 60 per cent in 1641 to about 5 per cent by the beginning of the nineteenth century. But much of this was achieved by landowners simply changing their religion and not by expropriation.
22. Peter Hart, *The IRA at War: 1916 to 1923* (Oxford: Oxford University Press, 2003), page 223.
23. Garvin *Preventing the Future*, page 2.
24. Cullen. *An Economic History*, page 176.
25. O'Duffy went on to found a quasi-fascist movement, the Blue Shirts, which later merged with the Cumann na nGaedheal party to form Fine Gael.
26. Quoted in McBrierty, *Ernest Thomas Sinton Walton*, page 36.

3. DARKENING CLOUDS

1. *De Valera's Irelands*, edited by Gabriel Doherty and Dermot Keogh (Cork: Mercier Press, 2003). Essay by Garret FitzGerald, 'Eamon De Valera, the Price of his Achievement', page 185.
2. See Doherty and Keogh, *De Valera's Irelands*, for accounts of Irish social development in this period.
3. See Garret FitzGerald, *Ireland in the World, Further Reflections* (Dublin: Liberties Press, 2005).
4. Gearoid O'Tuathaigh, 'Cultural Visions and the New State', in Doherty and Keogh, *De Valera's Irelands*, pages 166–84.
5. See map of Irish language penetration, FitzGerald, *Ireland in the World*.
6. McBrierty, *Ernest Thomas Sinton Walton*, page 39. De Valera in 1946 proposed a grant of £35,000 per annum for Trinity College, which by then was on the brink of insolvency.
7. O'Duffy was sacked as Garda Commissioner in 1933, about a year after the new government took office.
8. Dáil Reports, Vols 42–17, June 1932.
9. See debates on economic war and unemployment, Dáil Reports, Vols 50–4, 1934.
10. Ibid.
11. Cullen, *An Economic History*, page 178.
12. Dáil Reports Vol. 47, 3 May 1933.
13. Ibid.
14. Article 7 of the Anglo-Irish Treaty had specified that the Free State would provide 'in time of war or strained relations with a foreign power, such harbour and other facilities as the British Government may require' for defence purposes. An annex to the treaty listed four ports, three of which, Cobh, Berehaven and Lough Swilly, were located in the Free State area.
15. Quoted in Robert Fisk *In Time of War* (Gill & Macmillan, 1983), page 44.
16. The incident is vividly described in Sir Stanley Matthews' autobiography, *The Way it Was*, as follows: 'The beleaguered FA official left only to return some minutes later saying he had a direct order from Sir Neville Henderson, the British Ambassador in Berlin, that had been endorsed by the FA secretary Stanley Rous. We were told that the political situation between Great Britain and Germany was now so sensitive that it needed "only a spark to set Europe

alight". Faced with the knowledge of the direst consequences, we felt we had little choice in the matter, and reluctantly agreed to the request'.
17. Michael White and John Gribbin, *Einstein: A Life in Science*, 3rd edn (Free Press, 2005), page 206.
18. Rhodes *The Making of the Atomic Bomb*, pages 200–5.
19. Ibid.
20. Citation for award of Nobel Prize in chemistry, 1935.
21. Radium, which could only be produced in tiny quantities and which was prohibitively expensive, was also highly toxic and limited in its application. The Joliot Curies' discovery of a means of manufacturing artificial isotopes, especially using cobalt, was therefore of special significance to the more widespread availability of radioactive therapy in the treatment of cancer. Although the newer isotopes had much shorter half-lives than radium, they could later be produced by the ton, facilitating their far wider application in medical treatments.
22. *The World Set Free* envisaged the development of nuclear weapons and a nuclear war that would devastate civilisation in the 1950s. Wells' novel was apparently inspired by the earlier writings of Frederick Soddy on radio-activity and the atom.
23. Quoted in Rhodes, *The Making of the Atomic Bomb*, page 230.

4. WEAPON OF MASS DESTRUCTION

1. For a discussion on this aspect, see Rhodes, *The Making of the Atomic Bomb*, page 679.
2. The nucleus of a hydrogen atom contains one proton; the nucleus of a deuterium atom contains a proton and a neutron.
3. I.G. Farben was the German company which manufactured the gas used in concentration camps throughout the Holocaust.
4. Rhodes, *The Making of the Atomic Bomb*, page 303.
5. Einstein, as it turned out, was reluctant to write directly to the Queen, but drafted a letter to the Belgian ambassador to the US instead.
6. Letter dated 19 October 1939.
7. US Review Committee Report, 5 May 1941.
8. In February 1944, a British raid on the Norsk plant and the sinking of the Hydro ferry which was crossing to Germany with a consignment of heavy water effectively brought the German atomic research programme to an end.
9. A point noted by Einstein in his letter to Roosevelt in August 1939: 'Such bombs might very well prove too heavy for transportation by air.' In Germany, Heisenberg ultimately arrived at the same conclusion.
10. Quoted in Rhodes, *The Making of the Atomic Bomb*, page 325.
11. Ibid.
12. Rotblat was one of the only scientists to resign from the Manhattan Project on grounds of conscience.
13. Quoted in Preston, *Before the Fall Out*, page 158.
14. Tizard was a chemist and one of the priority concerns of his committee, the Committee on the Scientific Survey of Air Defence, was with development of radar systems for air-raid warnings.

15. Son of J.J. Thomson who had discovered the electron.
16. For an account of the events leading up to Churchill's appointment as PM, see Roy Jenkins, *Churchill* (London: Macmillan, 2001), pages 571–88.

5. 'FAT MAN' AND 'LITTLE BOY'

1. Report by MAUD Committee on the use of uranium as a source of power, March 1941.
2. The MAUD Committee then dissolved itself.
3. Plutonium occurs in nature, but only in minuscule amounts.
4. For a detailed description of early Anglo-American relations on the development of atomic weapons and the British programme, see Margaret Gowing, *Atomic Energy 1939–1945* (London: Macmillan, 1964).
5. Rhodes, *The Making of the Atomic Bomb*, page 388.
6. Quoted in ibid., page 440.
7. Up to that point, 46-year-old Groves' main claim to fame was that he had overseen the building of the Pentagon. See Peter Pringle and James Spigelman, *The Nuclear Barons* (New York: Holt, Rinehart & Winston, 1981).
8. Statements by the Prime Minister and Mr Churchill on the atomic bomb, 6 August 1945, appendix 1; Margaret Gowing, *Independence and Deterrence*.
9. Rhodes, *The Making of the Atomic Bomb*, page 502.
10. From Churchill's *The Second World War, the Gathering Storm*, Vol. 1, quoted in Robert Fisk, *In Time of War* (Dublin: Gill & Macmillan, 2004).
11. Dáil Reports Vol. 77, 2 Sept 1939.
12. Fisk, *In Time of War*, page 110.
13. The official estimates for Irish members of the British armed forces given after the war was between 28,000 and 48,000. However, estimates by historians of the period put the numbers at about 100,000.
14. Fisk, *In Time of War*, page 112.
15. For example Frank Aiken, latterly appointed Minister for Foreign Affairs, would almost definitely have opposed Irish entry into the war on the British side for ideological reasons.
16. O'Tuathaigh, *De Valera*, op cit page 70.
17. Roosevelt was not impressed. When the Irish Minister, Frank Aiken, visited the US in 1940 seeking American arms for Irish defence, Roosevelt was so enraged by his exposition of Ireland's neutrality policy that he pulled a tablecloth from a table set for tea, sending the cutlery flying in all directions. Ireland failed to secure the requested arms from the US. For account see Fisk, *In Time of War*.
18. Quoted in Fisk, *In Time of War*.
19. In fact, Craigavon was incensed when he came to know of the plan and accused the British government of betrayal. See Fisk, *In Time of War*.
20. De Valera's pleas to Britain and the US for defensive armaments had largely fallen on deaf ears.
21. Quoted in Jenkins *Churchill*, page 614.
22. Churchill to De Valera 8/12/1941. Quoted in Fisk, *In Time of War*, page 323. 'A Nation Once Again' was the anthem of Redmond's Irish Parliamentary

Party during the First World War and was unlikely to hold a positive resonance with De Valera.
23. Fisk, *In Tme of War*, page 167.
24. Both Germany and Britain developed tentative blueprints for an occupation of Ireland.
25. It also resulted in a remarkable public attack on De Valera and Irish neutrality in Churchill's victory radio broadcast at the end of the European war. De Valera's dignified response some days later, drawing attention to the rights of small nations to look after the preservation of their own interests, was hailed as a huge success in Ireland. In Britain, political and popular bitterness over De Valera's action in paying formal respects on the death of Hitler lingered for several generations.
26. See Rhodes, *The Making of the Atomic Bomb*, pages 625–40.
27. About $13.6bn in 2005 terms.
28. Rhodes, *The Making of the Atomic Bomb*, page 734.

6. BRITAIN'S BOMB FACTORY

1. Whitehaven's only previous historic claim to fame was that distinctive wooden pails made in the area received a mention in the tenth-century *Annals of the Four Masters*, written by monks in Co. Louth.
2. Some of the Whitehaven traders developed more respectable social connections with Britain's overseas colonies. The merchant George Gale married a widow, Mildred Washington, in Virginia in 1699. The Gales, including Mildred's three children, returned to live in Whitehaven. Within a year, Mildred had died and under the terms of her will, her children returned to America. Her son, Augustine Washington, was father of the first President of the United States, George Washington.
3. Department of Labour Survey Report, 1934. The baseline reports were undertaken on behalf of the Special Areas Commission in areas of particular social and economic deprivation.
4. Gowing, *Independence and Deterrence*, Vol. 2, page 393.
5. H.G. Davey, 'Reminiscences of an Atom Pioneer', unpublished memoir, available through National Archives, Kew.
6. Gowing *Independence and Deterrence*, Vol. 1, page 3. Keynes was sent off to negotiate a loan from the US, most of which was exhausted by late 1947.
7. Henry Tizard, scientific adviser to the Atlee government wrote in 1950: 'We are not a Great Power and never will be again. We are a Great Nation, but if we continue to behave like a Great Power we shall soon cease to be a Great Nation.' Quoted in Gowing, *Independence and Deterrence*, Vol. 1, page 229.
8. Letter Attlee to Truman, 25 Sept 1945. Full text reproduced in Gowing, *Independence and Deterrence*, Vol. 1, page 79.
9. Washington Declaration 15 Nov. 1945. See Gowing, *Independence and Deterrence*, Vol. 1 Appendix 4.
10. Britain as it turned out had sufficient stocks of Belgian uranium to supply its atomic programme until 1955.

11. Gowing, *Independence and Deterrence*, Vol. 1, page 110.
12. Nunn May was sentenced to ten years in prison, a sentence that most of the scientific community in the UK felt was unduly harsh, given his altruistic motivation.
13. See Richard Rhodes, *Dark Sun: the Making of the Hydrogen Bomb*, 2nd edn (New York: Simon & Schuster, 1996), page 127.
14. Quoted in Gowing, *Independence and Deterrence*, Vol. 1, page 164.
15. Throughout the wartime cabinet, Anderson had retained full responsibility for the atomic energy brief.
16. Groves said Anderson behaved 'as high-handedly as he had done in Ireland during the Black and Tan era'. Quoted in Pringle and Spigelman, *The Nuclear Barons*, page 81.
17. As Britain's official historian of the atomic defence programme, Margaret Gowing, put it: 'To those who recommend taking important matters out of party politics, the story of Sir John Anderson's responsibilities for atomic energy under the 1945 Labour Government is a warning rather than an encouragement.'
18. The Ministry of Defence was not founded until October 1946.
19. Gowing, *Independence and Deterrence*, Vol. 1, page 169.
20. About £970m in 2005 figures. The comparative cost is derived from dividing the cost index for 2005 by that for 1945 to calculate relative pound values. The overall capital cost of the British atomic programme in the period amounts to in excess of £2.2bn in 2005 terms.
21. Bevin's intemperate remark has been widely quoted in all accounts of the events leading to the decision to manufacture an independent atomic deterrent in the UK. In May 1944 the Americans had warned British scientists that the construction of a gaseous diffusion plant in Britain post-war, would amount to a breach of the Quebec Agreement Clause 4.
22. Gowing, *Independence and Deterrence*, Vol. 1, page 178.
23. Members of Gen 163 included Attlee, Bevin, Deputy Prime Minister Herbert Morrison (grandfather of Peter Mandelson), Addison, Secretary of State for Dominion Affairs, Minister for Defence Alexander Wilmot.
24. Gowing, *Independence and Deterrence*, Vol. 1, page 34.
25. HoC PQ 12 May 1951 Col 2,118.
26. Gowing, *Independence and Deterrence*, Vol. 1, page 166.
27. Ibid., page 176.
28. Ibid.
29. Cockcroft's team had built a prototype air-cooled design, BEPO, at Harwell. See Gowing, *Independence and Deterrence*, Vol. 2, pages 380–5.
30. Ibid., Vol. 1, page 79.
31. ORNL was established in Tennessee in 1943 as part of the Manhattan Project and housed a gaseous diffusion plant to separate U235 as well as a graphite reactor. Today, ORNL is the US primary nuclear research centre.
32. Davey, 'Reminiscences of an Atom Pioneer'.
33. The Russians' name for their first bomb test was 'First Lightning'. For a description of Russia's bomb test programme see Rhodes, *Dark Sun*, Chapter 19, pages 364ff.
34. Rhodes, *Dark Sun*, page 211.
35. Ibid, page 220.

36. Dáil Reports Vol. 113, 24 Nov. 1948.
37. Dáil Reports Vol. 112, 5 Aug. 1948.
38. The Irish government demanded Marshall Aid funds in grants rather than loans. In the end, they had to content themselves with mainly loan facilities. See Ferriter, *The Transformation of Ireland*, page 467.
39. Dáil Reports Vol. 115, 3 May 1949.
40. Dáil Reports Vol. 105, 25 March 1947.

7. PROLIFERATION

1. Quoted in Rhodes, *Dark Sun*, page 202.
2. Preston, *Before the Fall Out*, page 282.
3. Gowing, *Independence and Deterrence*, Vol. 1, page 10.
4. Preston *Before the Fall Out*, pages 282–4.
5. Bohr, for example, consistently reported any contacts made by the Russians, both during his time in the US and subsequently on his return to Copenhagen.
6. McLean and his colleague Burgess fled to the Soviet Union in 1951.
7. In a letter to his son John, then on active duty in the Korean War, President Eisenhower wrote: 'If the Soviets can convince prospective recruits that the worst possible penalty they would ever have to pay for exposure as spies would be a relatively short term in prison, then their blandishments and bribes would be much more effective ...' Quoted in Rhodes, *Dark Sun*, page 521.
8. Bruno Pontecorvo, who had been part of Fermi's team in Rome in the 1930s, worked at Los Alamos and became a naturalised Briton after the war, unaccountably defected to Russia with his wife and family in 1950.
9. The UN Commission was wound up in 1948 after failing to reach any agreement on a system of international control.
10. Jenkins, *Churchill*, page 786.
11. Acheson Lilienthal Report, Section 1, 'Background to the problem', Page 1.
12. Bevin delivered a paper to the cabinet in 1948 calling for Western European union.
13. The OEEC was the forerunner of the OECD and originally had eighteen members, including Ireland. The organisation was based in Paris.
14. Rhodes, *Dark Sun*, pages 299–301.
15. Gowing, *Independence and Deterrence*, Vol. 1, page 244.
16. Rhodes, *Dark Sun*, page 301.
17. Gowing, *Independence and Deterrence*, Vol. 1, page 265.
18. Ibid., page 272.
19. Ibid., page 289.
20. Ibid., pages 314–16.
21. Ibid.
22. The agreement with Union Minière, the Belgian company controlling the Shinkowolwe mines, and the Belgian government provided the entire output of the mine's cheap uranium supplies for Britain and the US up to 1956, from which Britain also netted a substantial dollar windfall due to the payment terms. See Gowing, *Independence and Deterrence*, Vol. 1, pages 366–8.
23. Ibid., page 421.

24. See *Guardian* 5 and 6 Aug 2005. Also BBC 'Newsnight'. UKAEA file documents. www.bbc.co.uk/newsnight. The 34 MW reactor at Dimona was capable of producing enough plutonium to make 1.2 bombs per annum. By the 1990s, it was estimated that Israel had a stockpile of about 200 nuclear weapons.
25. Gowing, *Independence and Deterrence*, Vol. 2, page 121.
26. Quoted in Jenkins, *Churchill*, page 873.
27. The IAEA was finally established in 1957.
28. *Irish Times*, 29 Feb. 1956 Headline: 'Plan for atomic pool for peaceful uses in Europe'.
29. *Irish Times*, 1 March 1956.
30. Mulcahy was responding to the launch of a review of education which advocated maintaining the status quo in a system of such acute educational disadvantage that nine out of ten Irish people failed to advance beyond primary education.
31. For analysis of this period see Garvin, *Preventing the Future*, Chapters 3 and 4.
32. Letter from John A. Costello to Professor Ernest Walton, 29 Feb. 1956. Walton family private papers.
33. *Irish Times*, 10 March 1956.
34. Address by Taoiseach John A Costello to inaugural meeting of the AEC, 15 April 1956. Walton Family Papers.
35. Equivalent to about 7 million euro in 2005 figures.
36. Ibid. Walton family papers.
37. Dáil Reports, Vol. 173, 4 March 1959.

8. HEAT AND LIGHT

1. See Walter C. Patterson, *Going Critical* (London: Paladin Books, 1985), page 9.
2. Investment in fast-breeder reactor development that used plutonium as fuel, a prototype of which had been built at Dounreay in Scotland, was finally abandoned as 'uneconomic' by the British government in 1988.
3. A prototype advanced gas reactor, WAGR, was built at Sellafield, a fast-breeder reactor at Dounreay in Scotland and a heavy water reactor at Winfrith in Dorset.
4. Pippa is an acronym for 'pressurised pile for producing power and plutonium'.
5. At a conference on nuclear power at Harwell in 1950, scientists and engineers considered a paper on the economics of nuclear power which emphasised the high capital costs of nuclear power stations.
6. Joule Memorial Lecture, 1951.
7. Gowing, *Independence and Deterrence*, Vol. 2, page 292.
8. Ibid.
9. Ibid.
10. For a discussion on the economics of early nuclear power production, see Gowing, *Independence and Deterrence*, Chapter 19, pages 262–301.
11. Quoted in Ibid., Vol. 1, page 432.
12. Jenkins, *Churchill*, page 892. The cabinet decision on the H-bomb was taken about a year before it was announced, in March 1954, with the aim of a first

test by 1958. The US and Soviet Union had already conducted H-bomb tests in 1952 and 1953, respectively.
13. Sir William Strath, head of the Cabinet Office War Plans Secretariat, led the team of civil servants who evaluated the consequences of a thermonuclear attack. Their report remained classified until its release in 2002.
14. Quoted in Michelle Tweena, 'Nuclear Energy – Rise, Fall and Resurrection', Center for International Climate and Environmental Research, March 2006.
15. *Irish Times*, 18 Oct. 1956.
16. Ibid.
17. Simpson, John, *The Independent Nuclear State: The United States, Britain and the Military Atom* (London: Macmillan, 1983), page 109. According to Simpson, the three power stations referred to appear to have been Hinkley Point, Dungeness and Sizewell.
18. Gowing, *Independence and Deterrence*, Vol. 2, page 26.
19. Ibid., Vol. 1, page 50.
20. Ibid., page 414.
21. Quoted in Gowing, *Independence and Deterrence*, Vol. 1, page 29.
22. Hansard, 2 Aug. 1946, Vol. 426 col. 1,371.
23. Quoted in Gowing, *Independence and Deterrence*, Vol. 2, page 137.
24. Ibid., page 133.
25. Although Hinton devised and implemented strict accounting procedures for construction at Windscale, Cockcroft was disconcerted to discover that expensive Portland stone was being used in the construction of the main administration building at Harwell.
26. Radioactivity was traditionally measured in 'curies', named after Marie Curie. A curie is the amount of radioactivity that decays at a particular rate and is usually expressed in curies per gram of material.
27. Gowing, *Independence and Deterrence*, Vol. 2, page 109.
28. Ibid., page 110.
29. Agencies and state bodies involved in the conduct and evaluation of Irish Sea experiments included the Royal Navy, Air Force, fisheries and marine biological laboratories, Cambridge Botany School and Oxford Agricultural School.
30. Gowing, *Independence and Deterrence*, Vol. 2, page 107.
31. The remarks were most recently quoted in an obituary for Dunster in the *Scotsman*, 1 June 2006.
32. Gowing, *Independence and Deterrence*, Vol. 2, page 278.
33. Arnold, Lorna, *Windscale 1957, Anatomy of a Nuclear Accident*, 2nd edn (London: Macmillan, 1995).
34. The reactors operated at 30 per cent less than their anticipated capacity.
35. Davey, 'Reminiscences'.

9. THE WINDSCALE FIRE 1957

1. Gowing, *Independence and Deterrence*, Vol. 2, Page 133.
2. Ibid., page 500.
3. See above, Chapter 8.

4. Britain and other European countries were largely dependent on the American monopoly of isotope production in the first decade after the war. However, the US imposed such onerous conditions on radioisotope exports and the isotopes themselves usually also required further refinement to make them suitable for use that there was general dissatisfaction and demand for development of European capacity. Britain's main production facility was located at Amersham. See Gowing, *Independence and Deterrence*, Vol. 2, Chapter 20 for an extensive account of the early development phase of radioisotope production in Britain.
5. See Arnold, *Windscale 1957*, page 133.
6. *Guardian* 6 Oct. 1987.
7. Quoted in Arnold, *Windscale 1957*, page 43.
8. Following the Suez crisis of 1956, Macmillan had succeeded Eden as Prime Minister. By mid-1957, the new PM had also taken over responsibility for the Atomic Energy portfolio.
9. In the 1950s, thyroid cancer could have proved fatal to its victims. The scientists' action in defining a threshold for contamination and imposing the milk ban undoubtedly saved lives. See Unscear 1993 Report, page 116.
10. All four Irish broadsheets, *Irish Times*, *Independent*, *Press* and *Cork Examiner*, carried the UKAEA quote in full in their morning editions of Saturday, 12 Oct. 1957.
11. J. Crabtree, 'The travel and diffusion of radioactive material emitted during the Windscale accident', *Journal of the Royal Meteorological Society* 85, 1959, pages 362–70. The filter system had been installed on public and private buildings throughout Europe in order to detect atmospheric radio-activity evidence of bomb tests in the Soviet Union.
12. A.C. Chamberlain, 'Emission of fission products and other activities during the accident to Windscale pile No. 1 in October 1957', UKAEA papers, 1981.
13. Magnox were the only British-designed reactors ever sold to countries outside the UK – to Italy and Japan. The Italian government approached the British embassy in Rome seeking further information on the fire. See Arnold, *Windscale 1957*, page 95.
14. Ibid., page 96
15. The Dublin Institute of Advanced Studies, with its School of Physics, had been established by Eamon de Valera at the outbreak of the Second World War, partly to provide a refuge for the quantum physicist, Erwin Schrödinger, who remained at the institute until 1956.
16. *Irish Independent*, 'Atom worry for English Farmers', 26 Oct. 1957.
17. Arnold, op cit page 62.
18. The academic engineers on the board of inquiry were Professor J.M. Kay, Professor of Engineering, Imperial College of Science and Technology and Professor Jack Diamond, Professor of Mechanical Engineering, Manchester University.
19. Quoted in Arnold, *Windscale 1957*, page 83.
20. Capenhurst, near Chester, was the site of the military programme's uranium enrichment plant, which went into operation in 1953.
21. *Guardian*, 25 Sept. 1987.

22. *Irish Press*, 19, 23, 26 and 29 Oct. 1957.
23. *Irish Press*, 19 Oct. 1957.
24. *Irish Press*, 23 Oct. 1957.
25. The Windscale fire was first mentioned in the Oireachtas during the Seanad debates on the Bill to establish the Nuclear Energy Board, Ireland's first regulatory agency, in 1971.
26. Quoted in Arnold, *Windscale 1957*, page 93.
27. Hansard 8 Dec 2004, Column 596W.
28. Davey, 'Reminiscences'.
29. The announcement that Windscale Pile No. 2 was also to be totally written off was made on 24 October 1958. Pile No. 2 had been temporarily shut down after the fire as a precautionary measure. Estimates for a Wigner release on the second pile, with all fuel removed, were of the order of £500,000. It was clear that the pile could not be restarted without risk of fire unless a Wigner release took place. Since plutonium from Calder Hall was coming on stream, the economic costs of keeping the second Windscale pile operational were judged prohibitive.
30. Quoted in Arnold, *Windscale 1957*, page 158.

10. IRELAND'S OPPORTUNITY

1. Devotion to the Child of Prague was particularly strong in Ireland, where the statue of the infant Jesus was, among other things, associated with influencing the weather. It was also common practice for brides-to-be to place the statue outside the house the night before their wedding, to bring blessings on their forthcoming marriage. The origins of the Child of Prague date back to Spain in the fourteenth century and it became particularly popular in Ireland from the mid-nineteenth century onwards. Electrified 'Sacred Heart' lamps were frequently given out free by the Electricity Supply Board as an inducement to householders to accept electrification during the Rural Electrification Scheme that began in the 1940s and continued over the next two decades.
2. Dáil Reports Vol. 160, 13 December 1956.
3. Dáil Reports Vol. 155, 14 March 1956.
4. Of which de Valera's 1943 St Patrick's Day broadcast, with its visions of 'comely maidens ... dancing at the crossroads' remains the most notorious example.
5. See Garvin, *Preventing the Future*, Chapter 4 for analysis of Ireland's education policy.
6. De Valera was elected President in April 1959 and Lemass became Taoiseach on 10 May 1959. De Valera finally retired from office in June 1973 and died on 29 August 1975 at the age of ninety-two years.
7. Dáil Reports Vol. 161, 20 March 1957.
8. For the seminal account of Lemass's life and times, see J. Horgan, *Sean Lemass: The Enigmatic Patriot* (Dublin: Gill and Macmillan, 1999).
9. Dáil Reports Vol. 219, 4 January 1966.
10. Dáil Reports Vol. 214, 10 February 1965. Answering Dáil questions on the meetings, Lemass said: 'The responsibility for ending the Partition of Ireland

rests on everybody who can make a contribution to it, including the British Government. I have on many occasions, however, expressed my own personal conviction that the primary aim must be to eliminate the barriers of suspicion and animosity which have divided the Irish people in the past so that unity can be founded on tolerance and goodwill. This is a task which has to be done in Ireland and by Irishmen'.
11. A decision of the Northern Irish Minister for Commerce, Brian Faulkner and the Republic's Minister for Transport and Power, Erskine Childers, to establish an expert committee to 'investigate the scope for agreement in cross-Border co-operation between the electricity supply systems, to advise on the economic and technical problems involved and to estimate broadly the savings in costs which might be achieved' was announced to the Dáil on 10 March 1965.
12. According to a report in the *Guardian*, Thursday, 8 Sept. 2005, referring to a visit by Lord Glentoran to the US in the 1950s, one American newspaper ran the headline 'Irish Royalty to visit Texas'.
13. Hansard 8 February 1955, Col. 1,720.
14. Lord Glentoran to Lord Salisbury, 2 June 1955, UKAEA Files AB 16/1732, National Archives, Kew.
15. Ibid. In a letter to his Chairman, Sir Edwin Plowden, early in July 1955, the Secretary of the UKAEA conceded: 'We have no absolute case for ruling out Northern Ireland altogether, now that we have gone to Dounreay.'
16. The Lord President of the Council, in charge of the Office of the Privy Council and a member of the cabinet, carried out other assignments as instructed by the Prime Minister. Salisbury was a well-known imperialist who had consistently opposed appeasement in the 1930s and served under three Prime Ministers in Britain – Churchill, Eden and Macmillan.
17. UKAEA Internal Correspondence AB 16/1732.
18. Springfields was the UKAEA's uranium manufacturing plant in Lancashire.
19. AB 16/1732.
20. Ibid. Sir Donald Perrot to Sir John Maud, 3 June 1955.
21. John Parkin to W.B. Lawson, Chief Civil Engineer, UKAEA, 11 November 1955, AB16/1732.
22. Dáil Éireann, Vol. 159, 26 July 1956.
23. *Irish Times*, 7 May 1958.
24. Dáil Reports Vol. 180, 5 April 1960.
25. The NPT finally came into force in May 1970.
26. Quoted in *Irish Times* report, 16 Nov. 1951.
27. Literal translation: Death/Life.
28. Dáil Reports, Vol. 173, 11 March 1959; Vol. 174, 8 April 1959.
29. Dáil Reports Vol. 182, 8 June 1960.
30. Of the first-generation nuclear programme in Britain, Berkeley and Bradwell came on line in 1962; Hunterston, Hinkley Point and Trawsfynydd in 1964; Dungeness and Sizewell in 1965; Oldbury in 1967 and Wylfa in 1970.
31. See Davey, 'Reminiscences'.
32. Simpson, *The Independent Nuclear State*, page 130.
33. Hesketh died in 2004, twenty-three years following his dismissal from the CEGB, during which time he remained convinced that the plutonium from

the civil reactors was transferred to the military plutonium stockpile in the UK and used in the US barter scheme. See London *Times*, Obituary for Ross Hesketh, 29 April 2004.
34. According to Simpson, *The Independent Nuclear State*, page 190: '... the only reactors potentially not involved in the transactions were those at Wylfa and possibly Oldbury. After April 1969, the material in used fuel rods remained the property of the CEGB and SSEB after reprocessing had taken place at Windscale but it was not until the end of 1970 that the last of the plutonium produced under the original arrangements was transferred to the United States. The plutonium reprocessed prior to April 1969 originated in fuel rods removed from reactors prior to mid-1968 and this, and the decision to open Bradwell to IAEA safeguards inspections in 1966, suggests that it was a product of no more than 30 months irradiation in the ten large reactors at Hunterston, Hinkley Point, Trawsfynndd, Dungeness and Sizewell and that it contained high percentages of Pu 239.'
35. Tom Wilkie, 'Old age can kill the Bomb', *New Scientist*, 16 Feb. 1984.
36. Gray et al. 'Discharges to the environment from the Sellafield Site, 1951–1992', *Journal of Radiological Protection*, March 1995.
37. The legislation setting up BNFL permitted 49 per cent of the company to be sold off subsequently without further recourse to parliament. In time, this became the basis for the proposed PPP – public–private partnership – future that BNFL anticipated in the 1990s.

11. THE PLUTONIUM ECONOMY

1. The 50 MW Calder Hall units compare with the twin 275 MW units for the first civil magnox station at Berkeley and the two 500 MW units for Hinkley Point.
2. Quoted in Patterson, *Going Critical*, page 11.
3. BNFL's reprocessing trade with Japan also ensured that nuclear power remained at the centre of diplomatic relations between the two countries from the 1970s onwards.
4. Pringle and Spigelman, *The Nuclear Barons*, pages 338–9.
5. Ghaddafi came to power in a military coup in Libya in September 1960, replacing the regime of King Idris, which had supplied oil to Europe through twenty-one different oil companies, including most of the major international companies. The Organisation of Petroleum Exporting Countries, OPEC, was founded at the Baghdad Conference, also in September 1960. Founding members were Iraq, Iran, Saudi Arabia, Kuwait and Venezuela. Within fifteen years, OPEC membership included practically all of the major oil-producing countries in the Middle East, Africa and South America.
6. See Tony Benn, *Against the Tide: Diaries 1973–77* (London: Arrow Books, 1990).
7. Pringle and Spigelman, *The Nuclear Barons*, page 365.
8. See Harold Bolter, *Inside Sellafield* (London: Quartet Books, 1996), pages 65–7.
9. The PIRA, 'Provos', emerged from a split in the IRA in 1969 into the Officials and Provisionals, with the latter favouring a purely military campaign to force the British out of Ireland.

10. *Irish Times*, 29 Dec. 2005. The first dinner with Callaghan was also attended by the then leader of Fianna Fáil, Jack Lynch.
11. FitzGerald, Garret, 'The 1974–5 Threat of a British Withdrawal from Northern Ireland', *Irish Studies in International Affairs*, RIA, 2005.
12. See Ferriter, *The Transformation of Ireland*, pages 622 ff.
13. Ibid.
14. Dáil Reports Vol. 303, 21 February 1978.
15. By mid-1971, the ESB had produced at least ten internal reports examining various aspects of nuclear power generation in Ireland, including studies on the economics of nuclear power and nuclear reactor safety.
16. The remaining two sites were on the Shannon Estuary in Co. Clare, and a site in Co. Waterford.
17. Dáil Reports, Vol. 255, 29 June 1971.
18. Ibid.
19. *Irish Times*, 1 Dec. 1973. The extent of local lobbying is also documented in state papers in this period.
20. The decision had been anticipated for some time in any case. Answering parliamentary questions on when the government would announce the nuclear project, Minister Peter Barry told opposition deputies: 'I hope to ask the government to make a decision within the next fortnight or three weeks.' See Dáil Reports Vol. 265, 17 May 1973.
21. In 2005 terms, £100m would equate to about 900 million euro.

12. WINDSCALE TO SELLAFIELD

1. In its first annual report for 1971, BNFL made no explicit reference to its role in the production of plutonium for military purposes. The report noted under Accounts that: 'Assets originally provided for Defence purposes and which the Company may in certain circumstances be required to use for such purposes had no value attributed to them on their transfer to the Company.'
2. Benn announced cabinet approval of BNFL's bid for 4,000 tonnes of spent fuel reprocessing contracts from Japan, and by extension his government's endorsement of the Thorp plan, in a written parliamentary reply on 12 March 1976.
3. See Bolter, *Inside Sellafield*, pages 177 ff for an account of this event.
4. Patterson, *Going Critical*, page 81.
5. Quoted in Bolter, *Inside Sellafield*, page 70.
6. A plant to provide a permanent solution to the management of this waste problem was finally commissioned in the 1980s.
7. Bolter, *Inside Sellafield*, page 72.
8. Patterson, *Going Critical*, page 86.
9. See Bolter, *Inside Sellafield*, page 250.
10. From the early 1950s, the scientists at Harwell researched a process for immobilising highly active liquid wastes in a glass matrix, a process known as vitrification. The project at Harwell was eventually abandoned and the UK bought in vitrification technology from the French.
11. Bolter, *Inside Sellafield*, page 95.

12. Quoted in Patterson, *Going Critical*, page 91.
13. See Chapter 8 above.
14. Reprocessing of spent nuclear fuels was impossible without a safe interim storage system for the highly active liquid wastes which it generated.
15. Peter Mitchell and Luis León Vintró, Department of Experimental Physics, UCD, 'What is the environmental Impact of Sellafield discharges?' (abstract), RIA 'Making Sense of Sellafield' Conference, 2003.
16. Pringle and Spigelman, *The Nuclear Barons*, page 253.
17. UNSCEAR 1993 Report, Annex B, page 115.

13. AN ACCEPTABLE RISK

1. The outcry over the test and growing public concern in the late 1950s and early 1960s generally about the scale of fallout from atmospheric tests spurred the first international agreement on an end to atmospheric testing of nuclear weapons in 1963, the Partial Test Ban Treaty. A Comprehensive Test Ban Treaty (CTBT) was finally adopted by the UN General Assembly on 10 September 1996.
2. C. O'Farrelly, *Nuclear Reactor Safety* RN/64 ESB 1971.
3. The China Syndrome refers to a theoretical scenario in which the reactor meltdown reaches the water table beneath, resulting in ever more violent hydrogen explosions. Ultimately, in theory, the molten core travels through the earth – from the US to China.
4. An independent containment was defined as one which would contain the worst hypothetical accident. In overground reactors, the containment system could not really be described as truly independent since it could not be guaranteed to preserve its integrity in the event of a reactor explosion.
5. NRS RN/64 ESB 1971 Recommendations.
6. In 1971, the ESB received an unsolicited bid for a partially underground nuclear power plant from the architect–engineering firm, Lahmeyer, who had operated as consultants to its hydroelectric complex at Turlough Hill. The proposal, which suggested building the power station in the side of a cliff, was rejected as it could not meet Teller's conditions for complete underground containment.
7. Ibid, Conclusions.
8. 'Containing the Incredible (Severe Accident and Containment Design Studies)', Paper by C. O'Farrelly presented to Seminar on 'Severe Accidents in Light Water reactors', CEC Brussels 10–12 November 1986.
9. Ibid.
10. *Irish Times*, 'ESB Defers Plan for Nuclear Power', 4 Oct. 1975. The paper reported the proposed costs of the project had now risen to £200m. The project was being deferred 'because of a reduction in the demand for electricity and because of the current difficulties in getting suitable long-term funds for the huge investment involved'.
11. Nuclear Energy Board, 2nd Annual Report, 1975.
12. NEB, 3rd Annual Report, 1976.
13. See Chapter 10 above.

14. Dáil Reports, Vol. 338, 2 Nov. 1982.
15. Dáil Reports, Vol. 323, 4 Nov. 1980.
16. According to CND, thirty local authorities in the Republic and Northern Ireland declared themselves nuclear-free zones between 1981 and 1988.
17. It has become commonplace, if entirely inaccurate, in parliamentary debates for public representatives to claim that the public protests succeeded in halting the Carnsore nuclear project. For example, Dáil Reports Vol. 621, 15 June 2006: 'We need only remember the attempt by the founder of Fianna Fáil's partner in Government, Des O'Malley, when he was a Fianna Fáil Minister to ram through a nuclear power station at Carnsore Point in Wexford. It was only because of the democratic voice of the people of Ireland who took to the streets in protest that this action was prevented', Aengus O'Snodaigh, TD, Sinn Féin; or Seanad Reports Vol. 168, 10 Oct. 2001, Labour Senator Joe Costello: 'This country decided against nuclear power nearly a quarter of a century ago. The people took to the streets and to the fields, roads, avenues and little laneways of Wexford, down to Carnsore Point. The end result was that despite a determined movement by the Government of the day, the decision was taken and that decision was never reversed.'

14. THE LEGACY OF THREE MILE ISLAND

1. For a detailed and comprehensive historical account of the TMI accident see J. Samuel Walker, *Three Mile Island: A Nuclear Crisis in Historical Perspective* (University of California Press, March 2004).
2. Pringle and Spigelmann, *The Nuclear Barons*, page 420.
3. No new reactors were ordered in the US post-TMI, although existing orders on the books at the time of the accident were completed.
4. The Three Mile Island station was located about 16km south east of Harrisburg, the state capital of Pennsylvania.
5. Quoted in Pringle and Spigelmann, *The Nuclear Barons*, page 438.
6. C. O'Farrelly, *Containing the Incredible*, Chapter 12.
7. Dáil Reports, Vol. 298, 29 March 1977.
8. At the Fianna Fáil Ard Fheis (National Conference) in January 1978, O'Malley had rounded on the anti-nuclear movement and threatened local opposition to the project in Wexford that it might be moved elsewhere – to Sligo or Mayo – if it was rejected locally.
9. Dáil Reports Vol. 315, 21 June 1979.
10. The state formally divested itself of the INPC only in 2001, when the company was sold. NORA, the National Oil Reserve Association, remains in state ownership.
11. A reference to Sean McBride, son of the legendary Maud Gonne and Major John McBride who was shot for his part in the 1916 Easter Rising. McBride was himself an IRA veteran of the War of Independence and subsequent IRA leader who later became Tanaiste and leader of the (by now defunct) Clann na Poblachta party in the first inter-party government 1948–51. McBride was co-winner of the Nobel Peace Prize in 1974 and was awarded the Lenin Peace Prize in 1977 in recognition of his work as a co-founder of Amnesty

International. However, in Ireland he remained a controversial and shadowy left-wing political figure to the end of his life.
12. See Dáil Reports Vol. 303, 21 Feb. 1978: Private Members' Business – Energy Policy: Motion in the name of the Labour Party.
13. Dáil Reports Vol. 309, 23 November 1978, Electricity (Supply) (Amendment) Bill, 1978: Second and subsequent stages.
14. Until the publication of the Green Paper on Energy on 1 October 2006, O'Malley's Green Paper was the first and only government policy on energy ever published.
15. Dáil Reports Vol. 309, 23 November 1978.
16. Firth was master of a trading vessel, the *Cumbria*, on the Irish Sea run in 1916. On a trip to the Isle of Man the *Cumbria* was confronted with a German submarine and engaged in a firefight. The flashes of the guns could be seen from the Co. Down coast, but Firth outwitted the submarine by zigzagging his ship to evade shells, while responding with his own fire. Eventually, the submarine backed off. Firth was awarded a medal which, he later claimed, his children lost while playing with it.
17. Dáil Reports Vol. 313, 2 May 1979: Parliamentary Questions.
18. Dáil Reports Vol. 315, 27 June 1979 Adjournment Debate – Wylfa Nuclear Power Station.
19. *Irish Press*, 23 June 1979.
20. Dáil Reports Vol. 315, 26 June 1979.
21. Quoted in T. Ryle Dwyer, *Haughey's Forty Years of Controversy* (Mercier Press, 1992, 2005), page 78.
22. NEB Annual Report for 1981 noted: 'Because of the Government's continuing low priority for the nuclear option for power generation in Ireland, the Board's activities in this field have been confined to maintaining a watching brief on developments in reactor safety and radioactive waste disposal.'
23. Dáil Reports Vol. 322, 12 June 1980, Parliamentary Questions.
24. Throughout his early parliamentary career, Haughey made few, if any, contributions to debates on energy issues. As Minister for Justice in the early 1960s, Haughey had strongly defended the Gardaí against an accusation they had set unmuzzled Alsatian dogs on Dr Noel Browne during an Irish Committee for Nuclear Disarmament demonstration at the US embassy in Dublin in 1962. The Minister for Justice gave no intimation of having even the slightest sympathy with Dr Browne's cause. Dáil Éireann Reports, Volume 197, 31 October 1962. Commenting on Haughey's record in May 1986, Labour leader Dick Spring stated: 'Certainly during the period of Fianna Fáil Government between 1977 and 1981, during which time Deputy Haughey was both Minister and for a certain period of time Taoiseach, I cannot find any record of any statement of dissent by him in either of those roles in relation to the course of action which the then Minister, Deputy O'Malley, was proposing in relation to a nuclear power station at Carnsore Point in County Wexford.' Reports, Vol. 365, 6 May 1986.
25. By 1979, the Northern authorities had advised the Republic's government that they were no longer in a position to repair further breaches of the interconnector.

15. ATOMIC VILLAGE

1. Davey, 'Reminiscences'.
2. Most of the radioactivity released during the TMI accident comprised noble gases, deliberately released to reduce pressure in the containment building.
3. See Arnold, *Windscale 1957*, pages 146–7.
4. See *Britain's Nuclear Nightmare*, by James Cutler and Rob Edwards (London: Penguin, 1988) for a full account of the making of the 'Nuclear Laundry' programme.
5. The *Guardian* and Independent Television had carried reports of the workers' deaths, one from leukaemia and the other from myleoma.
6. E. A. Clough, 'BNFL Radiation Mortality Study', *Journal of the Society for Radiological Protection*, Autumn 1983. The main purpose of the article was to provide comparisons of mortality among the male workforce at Sellafield to the end of 1975, distinguishing between three employee categories and between radiation and non-radiation workers.
7. Smith, P.G. and Douglas, A.J. 'Mortality of workers at the Sellafield plant of British Nuclear Fuels', *BMJ*, Clin. Res. Ed., Vol. 293, No. 6551, 1986.
8. Previously, BNFL and other nuclear companies vigorously contested all claims in the courts. The compensation scheme was later extended to cancers that did not result in death and, in time, was joined by other industry companies including the UKAEA.
9. As UKAEA and BNFL 1970s chairman Sir John Hill, had observed: 'The public doesn't and cannot be expected to understand the issues of nuclear power in other than the broadest terms.'
10. Figures quoted in Michelle Tweena, 'Nuclear Energy – Rise, Fall and Resurrection', Center for International Climate and Environmental Research, March 2006.
11. Sizewell B was the last reactor built in the UK in the twentieth century.
12. Sir Douglas Black's 1980 study of the relationship between social class, corresponding lifestyle habits, poverty, social exclusion and health sought to explain why the lower socio-economic groups had benefited less from the NHS than higher-income groups in society. His report was effectively rejected by the Thatcher government of the day as too expensive to implement. Much of Black's earlier work was concerned with the effects of dehydration on human health and mortality rates. Following a long and distinguished career, Sir Douglas Black died in 2002 at the age of eighty-nine.
13. Sir Richard Doll discovered the link between cigarette smoking and lung cancer. For over twenty years, he authored and contributed to a wide-ranging body of work on the nuclear installations' cancer clusters phenomenon. Sir Richard Doll died in 2005.
14. See *Daily Telegraph*, 11 July 2006 for latest report confirming validity of Kinlen's theory.
15. See Bolter, *Inside Sellafield*, pages 100–6.
16. See ibid., page 186. The courts eventually accepted £36,000 in payment of the fine and BNFL did not pursue costs against Greenpeace.
17. The Maze hunger strikes came to an end on 3 October 1981.
18. Dáil Reports, Vol. 350, 8 May 1984.

19. Dáil Reports, Vol. 350, 8 May 1984.
20. *British Medical Journal*, 12 November 1983, Vol. 287.
21. See NEB annual reports 1980 and following years.

16. CLOSE DOWN SELLAFIELD

1. There is a vast literature on the Chernobyl accident. For factual data on the causes of the accident and its aftermath and long-term environmental and health effects readers are referred to the various accounts of UN bodies, including the Chernobyl Forum, all of which material is accessible via the internet.
2. *Newsweek*, 9 June 1986.
3. Reported in *Irish Times*, Thursday, 1 May 1986.
4. *Irish Times*, 29 Apr. 1986.
5. *Irish Times* 30 Apr. 1986.
6. Bolter, *Inside Sellafield*, pages 221–2.
7. Seanad Éireann, Vol 112, 1 May 1986.
8. *Sunday Press*, 11 May 1986.
9. Ibid.
10. Dáil Éireann Vol. 365, 6 May 1986, col. 2,858.
11. Fianna Fáil's first 'close Sellafield' motion was put down on 11 March 1986. See Dáil Reports, Vol. 364.
12. Dáil Reports, Vol. 370, 26 November 1986.
13. The discrepancy was discovered by the UK scientist Dr Derek Jakeman, who had worked at Windscale in the 1950s and thereafter spent several years in Switzerland. On his return to England in 1985, Jakeman concluded that the data for pre-1957 emissions could not possibly be right. However, he was frustrated in his quest to extract further details from BNFL's Director of Health and Safety, Peter Mummery, who argued that it was impossible to provide accurate figures as the data had not been properly recorded at the time. Nevertheless, BNFL was eventually forced to concede that the 440 gm figure initially supplied to Black was well below the real level of emissions and the new data was included in the first report of COMARE in 1986. Having considered the new data, Dr Black found no reason to alter his original conclusions. Elsewhere, particularly in Ireland and among anti-nuclear lobby groups, the data discrepancy was held as discrediting the Black report.
14. Dáil Reports, Vol. 370, 3 December 1986.
15. Dáil Reports, Vol. 366, 7 May 1986. 'The kernel of the issue is that Britain could be taken to the European Court for being in breach of EC regulations causing a threat to a sovereign state bordering it. There are very serious grounds for believing that that could be done because of the risk which it is putting before this nation,' Fianna Fáil Deputy G. Brady proposed.
16. Dáil Reports Vol. 370, 3 December 1987.
17. Dáil Reports Vol. 373, 3 June 1987.
18. Bolter, *Inside Sellafield*, page 193.
19. Seanad Reports, Vol. 118, 20 January 1988.

17. BATTLE STATIONS

1. A large cohort of the workforce on the dam for the lake had been recruited from Co. Cork, many of whom later settled in the area, boosting the local population, and for whose religious convenience a local disused tannery was later converted into a Catholic church.
2. RTÉ Transcript, 'Morning Ireland', 19 Jan. 1988.
3. NEB Files, Proposed Tests at Trawsfynnydd Nuclear Power Station: Notes on Meeting with Minister, Tuesday 19 Jan. 1988.
4. Seanad Reports, Vol. 118, 20 January 1988.
5. NEB Files, Letter to the Board dated 20 Jan. 1988.
6. NEB Files, Letter to Chairman dated 22 Jan. 1988.
7. Department of Energy Statement 27 Jan. 1988.
8. Dáil Reports Vol. 377, 28 Jan. 1988.
9. CEGB press release, 3 Feb. 1990.
10. See Chapter 9, above.
11. Dáil Reports Vol. 398, 10 May 1990.
12. Dáil Reports Vol. 383, 20 Oct 1988.
13. Dáil Reports Vol. 390, 18 May 1989.
14. Dáil Reports Vol. 388, 18 April 1989. This policy, which was maintained by successive Irish governments, was particularly curious since in the event of a nuclear accident in Britain that affected the economic interests, health or property of Irish citizens living in Ireland, they would be forced to pursue remediative legal action individually through the courts whereas the conventions provided for automatic payments.
15. See Ryle Dwyer, op cit page 159.
16. Quoted in Ryle Dwyer, op cit page 207.
17. See Mahon Tribunal Report, 2006.

18. THORP

1. Gardner et al., 'Results of case-control study of leukaemia and lymphoma among young people near Sellafield nuclear plant in West Cumbria', *BMJ*, 17 February 1990.
2. Parker et al., *BMJ*, 1993.
3. Quoted in Bolter, *Inside Sellafield*, page 157.
4. See COMARE Reports 4 and 7, available at www.comare.org.uk.
5. Greenpeace UK, press release, 8 Oct. 1993.
6. See Chapter 10 above.
7. Quoted in *Irish Times* 'Bono rolls out the barrel at Sellafield', 22 June 1992.
8. *Sunday Times*, 21 June 1993.
9. Friends of the Earth later referred to John Gummer in a press release (11 Dec. 1995) as 'probably one of the greenest Environment Secretaries we've ever had'.
10. Hansard HoC debates, 13 Dec. 1993, Col. 1,094; Also Dáil Reports Vol. 436, 16 Dec 1993.
11. See Bolter, *Inside Sellafield*, page 84.

12. Hansard, ibid. The section in Brian Cowen's Dáil speech referring to the number of objections was clearly directly lifted from Gummer's House of Commons statement, indicating the flow of information between the two governments on the issue, as follows: 'Of the 42,500 responses received in London 63 per cent were opposed to the operation of THORP. More than 29 per cent of these respondents called for a hearing or public inquiry. This demand included 85 local authorities in the UK as well as the Isle of Man Government, although the two local authorities with direct responsibility for the Sellafield area – Cumbria County Council and Copeland District Council – did not consider a hearing to be necessary.' Cowen, December 14. 'Altogether, 42,500 individual responses were received, 63 per cent of which were opposed to the operation of Thorp. Overall, 29 per cent of respondents called for a hearing or public inquiry. That total included 85 local authorities, although Cumbria county council and Copeland district council – the two authorities with direct responsibility for the Sellafield area – did not consider a hearing to be necessary.' Gummer, 13 December 1993.
13. FoE press release, 23 Oct. 1998.
14. See Chapter 10 above.
15. The EC Opinion had been issued on 30 April 1992.
16. Following the assassination of Ambassador Christopher Ewart Biggs, vehicles used by British ambassadors to Ireland were specially armoured to resist explosions and firearms attacks.
17. Hansard HoC debates 15 Dec. 1993, Cols 1,094–1,104.
18. Dáil Reports Vol. 436, 14 Dec 1993.
19. Dáil Reports Vol. 439, 1 March 1994.
20. Lowe, V.P. 'Radiation and the Birds at Ravenglass', *Journal of Environment Pollution* 1991 Vol. 70 (1) 1–26.
21. Jay M. Gould and Benjamin A. Goldman, *Deadly Deceit, Low-Level Radiation, High-Level Coverup* (New York: Four Walls Eight Windows, May 1990).
22. Dáil Reports Vol. 439, 1 March 1994.
23. Dáil Reports Vol. 437, 26 Jan 1994.
24. Dáil Reports Vol. 439, 1 March 1994.
25. Dáil Reports Vol. 444, 22 June 1994.
26. The four Dundalk residents were Constance Short, Mary Kavanagh, Mark Dearey and Ollan Herr.

19. A POLITICAL BONE OF CONTENTION

1. *Irish Times*, 18 March 1997.
2. The Flowers report also counselled against the build up of unwarranted plutonium reserves in Britain. However, at the time of the report, a much greater expansion in nuclear power in Britain was envisaged than subsequently occurred.
3. The NIREX list was finally published only in 2005, following a Freedom of Information Act request by Friends of the Earth.
4. Two Irish MEPs, Patricia McKenna of the Green Party (Dublin) and Jim Fitzsimons of Fianna Fáil (Leinster), also directly gave evidence to the inquiry.

5. *Irish Times*, 11 March 1997.
6. FoE press release, 17 March 1997.
7. Including in a letter to the *Irish Times*, 18 Feb. 1997.
8. BNFL press release, 18 March 1997.
9. Germany was a notable exception. The German government had decreed, without public consultation, that high-level waste returned from reprocessing in Britain and France would be stored at Gorleben, and the site became the focal point for mass demonstrations against nuclear waste rail transports throughout the 1990s.
10. A second bilateral draft was presented to the Irish government in 1997.
11. The UK Department of Energy was subsumed within the Department of Trade and Industry, DTI, in 1989.
12. See Chapter 11 above.
13. *Irish Times*, 15 May 1986.
14. Seanad Reports, Vol. 149, 30 Oct 1996.
15. The BIIPB was founded in 1990.
16. See Seanad Debates, Vol. 180, 25 May 2005. Commenting on a visit to Sellafield, Fine Gael senator James Bannon noted: 'I spent two days visiting the plant and returned to Ireland with a great sense of fear.'
17. See Harbours Act, 1996. Also *Irish Times*, 28 May 1996.
18. Specifically Germany, Italy, Japan, the Netherlands and Switzerland. 'It was pointed out by the Governments concerned that they could not intervene directly in contractual agreements entered into with British Nuclear Fuels,' Stagg conceded subsequently.
19. The commitment to take legal action against the British government had been included in the coalition's plan for its period in office: 'A Programme for Renewal'.
20. Quoted in *Irish Times*, 15 Jan. 1996.
21. Shortly afterwards, BNFL accepted an invitation from Dermot Ahern TD to participate in a questions and answers debate in Dundalk. The meeting took place in April 1997. BNFL had also attended a conference of the General Council of County Councils in Galway in 1995.
22. As reported in the *Irish Times*, 29 Sept. 1997.
23. *Irish Times*, 16 Jan. 1997.
24. *Irish Times*, 29 Sept. 1997.
25. See Fred Pearce, 'Greenpeace: Storm-Tossed on the High Seas', *Green Globe Yearbook, 1996*.
26. Battersby, Eileen, 'Across the bitter sea', *Irish Times*, 23 Jan. 1997.
27. Ibid.
28. Transcript LMFM, 'Loose Talk', 6 Oct. 1998.
29. Ibid. Two of the main industries in Dundalk at that time were cigarette manufacture and a brewery.
30. See Dáil Reports Vol. 481. Private Members Debate: 'Case against BNFL', 14–15 October 1997.
31. Transcript 98 FM News, 17 May 1997.

20. FRIENDS AND NEIGHBOURS

1. Dáil Reports Vol. 493, 24 June 1998.
2. Dáil Reports Vol. 499, 3 Feb. 1999.
3. Dáil Reports Vol. 543, 6 Nov. 2001.
4. Reuters, 15 July 1999.
5. See (London) *Independent*, 21 July 1999.
6. The *Independent* story was published on 14 Sept. 1999.
7. On 30 March 2000 the British government announced that the BNFL privatisation would be postponed until the end of 2002.
8. *Irish Times*, 11 Feb. 1997.
9. Dáil Reports Vol. 514, 22 Feb. 2000.
10. Jo Moore was forced to resign following a leaked memo on councillors' expenses that she had circulated on 11 September 2001; a good day, she suggested, to bury bad news.
11. Michael McDowell was appointed Attorney General in July 1999. See Dáil Reports Vol. 514, 22 Feb. 2000.
12. G. Dean et al., 'Investigation of a cluster of children with Down's syndrome born to mothers who had attended a school in Dundalk, Ireland', *Occupational and Environmental Medicine* 2000, 57: 793–804.
13. RPII press release, 10 Apr. 2001.
14. See Hansard 21 June 2000, Column 95WH.
15. DEFRA, the UK Department of Environment, Forestry and Rural Affairs, appointed independent consultants Arthur D Little (ADL) in April 2001 to review the economic case for operating SMP. The ADL report concluded there was a robust economic case to support SMP. Their report provided the basis for the fifth and final public consultation on the MOX plant project.
16. Dáil Reports Vol. 541, 4 Oct. 2001.
17. Transcript RTÉ 'Prime Time', 4 Oct. 2001.
18. Dáil Reports Vol. 542, 23 and 24 Oct. 2001.
19. Greenpeace, CND and Friends of the Earth quit the stakeholder dialogue process following the decision to authorise the Sellafield MOX plant.
20. Hansard HC 1122-v Uncorrected transcript, 19 June 2006.
21. *Irish Independent*, 17 Feb. 2003.
22. BNFL press statement, 10 Oct. 2001.

21. PYRRHIC VICTORIES

1. Transcript RTÉ Radio 1, 'Marian Finucane Show', 26 Sept. 2001.
2. *Star* (Irish edition) 9 Oct. 2001.
3. Transcript RTE Radio 1, 'Marian Finucane Show', 7 Nov. 2001.
4. The nearest point in the Republic of Ireland to Sellafield, Clogher Head, in Co. Louth is in fact 121 miles from Sellafield.
5. LMFM. 'Loose Talk'. Interview with Paul Maguire, 9 Oct. 2001.
6. Transcript RTE Radio 1 'Morning Ireland', 17 Dec. 2001.
7. RPII Report, November 2001.
8. Dáil Reports Joint Committee on Environment, 11 May 2005.

9. BNFL and the RPII finally met in Dublin on 14 January 2002 to review the new data on HAL storage safety.
10. BNFL press statement, 25 Jan. 2002.
11. See RPII: 'Report on visit to Sellafield, April 2005'. On page 15, the RPII noted: 'It is not possible for the NII to share detailed information with the Institute to allow it to make its own assessment of the likely consequences for Ireland of a terrorist attack ... While accepting fully the need to protect sensitive information about plant security, the lack of an established framework for assessing the adequacy of threat assessments and security arrangements remains a significant concern.'
12. Hansard N.I. Assembly debates, 4 Dec. 2001.
13. RPII 2000 figures, quoted in Mitchell & Vintro, 'Making Sense of Sellafield' (abstract), RIA September 2002.
14. Ibid.
15. NORM is discharged as a result of phosphate fertiliser production, although such discharges have been reduced since the 1990s, and from the extraction of oil and gas from the continental shelf in the North Sea, mainly in the Norwegian and U.K sectors. NORM accumulates as scale inside pipework and valves at offshore oil and gas production platforms. It also gathers as sludge in separator tanks and other vessels. It is discharged in 'produced water' and its radionuclides of radium-226 and Ra-228 and Pb-210 (lead) become available in concentrated form for consumption by marine biota. Apart from oil and gas production, NORM also arises in peat burning, bauxite production and the manufacture of cement.
16. Pilot study for the update of the MARINA Project on the radiological exposure of the European Community from radioactivity in North European marine waters, European Commission, December 1999.
17. ITLOS 2001 Order; 3 Dec. 2001.
18. *Irish Times*, 'Irish debate on Sellafield "dishonest"', 29 May 2003.
19. Transcript, Radio Ulster, 10 June 2003.
20. London *Independent*, 15 June 2005.
21. Ibid.
22. 'Chernobyl's Legacy: Health, Environmental and Socio-Economic Impacts', Chernobyl Forum, September 2005. The 600-page UN report incorporates the work of some 400 scientists, economists and health experts. It concludes that as of mid-2005 fewer than fifty persons had died directly as a result of the disaster and nine children from 2,000 cases of thyroid cancer directly related to its radioactive fallout. It estimates that the ultimate death toll will be of the order of 4,000 and that mental health problems are the largest public health problem created by the accident. The forum is made up of eight UN specialised agencies, including the International Atomic Energy Agency (IAEA), World Health Organisation (WHO), United Nations Development Programme (UNDP), Food and Agriculture Organisation (FAO), United Nations Environment Programme (UNEP), United Nations Office for the Coordination of Humanitarian Affairs (UN-OCHA), United Nations Scientific Committee on the Effects of Atomic Radiation (UNSCEAR), and the World Bank, as well as the governments of Belarus, the Russian Federation and Ukraine.

23. Hansard 10 July 2002, Column 991W.
24. *Irish Independent*, 8 Dec. 2004.
25. Dáil Reports, Parliamentary Questions, 29 Nov 2006.
26. Dept. of Environment, press release, 18 Jan. 2006.
27. Dáil Reports, Transcript, 22 Feb 2007.
28. See NDP 2007–2013, Jan. 2007.
29. Joint Convention on the Safety of Spent Fuel Management and on the Safety of Radioactive Waste Management. National Report by Ireland, 27 Oct 2006
30. Council Directive 2003/122/Euratom of 22 December 2003 on the control of high-activity sealed radioactive sources and orphan sources.
31. Dáil Reports, Joint Committee on Environment and Local Government, 7 Jan. 2004.
32. Ibid.
33. Ibid.
34. Radiological Protection Institute of Ireland, Annual Report and Accounts 2005, page 9.
35. Johnson, C., et al., 'A study of the movement of radioactive material released during the Windscale fire in October 1957 using ERA40 data', *Atmospheric Environment* (2007), oi:10.1016/j.atmosenv.2006.11.058. See also Gardner, Wakeford et al., 'Atmospheric Emissions from the Windscale Accident of 1957', available at www.sciencedirect.com.
36. 'Retrospective Search for Evidence of the 1957 Windscale Fire in NE Ireland Using 129I and Other Long-Lived Nuclides', D. Gallagher, E.J. McGee, P.I. Mitchell, V. Alfimov, A. Aldahan, and G. Possnert, *Journal of Environmental Science and Technology* Vol. 39 No. 9, 1 May 2005, pp. 2,927–35. Also, *Irish Times* report, 31 March 2005.
37. Ibid.

Select Bibliography

Books

Arnold, Lorna, *Windscale 1957, Anatomy of a Nuclear Accident* (London: Macmillan, 1992 and 1995).
Bolter, Harold, *Inside Sellafield* (London: Quartet Books, 1996).
Cathcart, Brian, *The Fly in the Cathedral*, 2nd edn (London: Penguin, 2005)
Cullen, L.M. *An Economic History of Ireland since 1660* (London: B.T. Batsford, 1972).
Cutler, James and Edwards, Rob, *Britain's Nuclear Nightmare* (London: Penguin, 1988).
Davey, H.G. 'Reminiscences of an Atom Pioneer', unpublished memoir (London: UKAEA 1958).
Doherty, Gabriel & Keogh, Dermot (eds), *De Valera's Irelands* (Cork: Mercier Press, 2003).
Ferriter, Diarmaid, *The Transformation of Ireland 1900–2000* (London: Profile Books, 2004).
FitzGerald, Garret, *Ireland in the World: Further Reflections* (Dublin: Liberties Press, 2005).
Fisk, Robert, *In Time of War* (Dublin: Gill & Macmillan, 1983).
Garvin, Tom, *Preventing the Future; Why was Ireland so Poor for so Long?* (Dublin: Gill & Macmillan, 2004).
Gowing, Margaret, *Independence and Deterrence; Britain and Atomic Energy 1945–52 Vols 1 & 2* (London: Macmillan, 1972).
Gowing, Margaret, *Atomic Energy 1939–1945* (London: Macmillan, 1964).
Hart, Peter, *The IRA at War: 1916 to 1923* (Oxford: Oxford University Press, 2003).
Horgan, J., *Sean Lemass: The Enigmatic Patriot* (Dublin: Gill & Macmillan, 1999).
Jenkins, Roy, *Churchill* (London: Macmillan, 2001).
McBrierty, Vincent J., *Ernest Thomas Sinton Walton 1903–1995: The Irish Scientist* (Dublin: Trinity College Dublin Press, 2003).
O'Carroll, J.P & Murphy John A. (eds), *Eamon De Valera and his Times* (Cork: Cork University Press, 1985).

Patterson, Walter C., *Going Critical* (London: Paladin Books 1985).
Preston, Diana, *Before the Fall Out: The Human Chain Reaction from Marie Curie to Hiroshima* (New York: Doubleday, 2005).
Pringle, Peter and Spigelman, James, *The Nuclear Barons* (New York: Holt, Rinehart & Winston, 1981).
Rhodes, Richard, *The Making of the Atomic Bomb* (New York: Touchstone/Simon & Schuster, 1987; 1995).
Rhodes, Richard, *Dark Sun: the Making of the Hydrogen Bomb*, 2nd edn (New York: Simon & Schuster, 1996).
Simpson, John, *The Independent Nuclear State: The United States, Britain and the Military Atom* (London: Macmillan, 1983).
White, Michael & Gribbin, John, *Einstein – A Life in Science,* 3rd edn (London: Free Press, 2005).

Journals and Periodicals

Chamberlain, A.C., 'Emission of fission products and other activities during the accident to Windscale pile No.1 in October 1957', UKAEA papers, 1981.
Clough, E.A., 'BNFL radiation mortality study', *Journal of the Society for Radiological Protection,* Autumn 1983.
Crabtree, J., 'The travel and diffusion of radioactive material emitted during the Windscale accident', *Journal of the Royal Meteorological Society* 85, 1959.
FitzGerald, Garret, 'The 1974–5 threat of a British withdrawal from Northern Ireland', *Irish Studies in International Affairs*, RIA, 2005.
Gardner et al., 'Results of case-control study of leukaemia and lymphoma among young people near Sellafield nuclear plant in West Cumbria', *BMJ*, 17 February 1990.
Gardner, Wakeford et al., 'Atmospheric Emissions from the Windscale Accident of 1957', available at www.sciencedirect.com.
Gallagher et al., 'Retrospective search for evidence of the 1957 Windscale fire in NE Ireland Using 129 I and Other Long-Lived Nuclides', *Journal of Environmental Science and Technology*, Vol. 39, No. 9, 1 May 2005.
Gray et al., 'Discharges to the environment from the Sellafield Site, 1951–1992', *Journal of Radiological Protection*, March 1995.
Johnson, C. et al., 'A study of the movement of radioactive material released during the Windscale fire in October 1957 using ERA40 data', *Atmospheric Environment* (2007), oi:10.1016/j.atmosenv.2006.11.058.

Lowe, V.P. 'Radiation and the birds at Ravenglass', *Journal of Environment Pollution*, Vol. 70 (1) 1991.

McLaughlin et al., 'Paternal radiation exposure and leukaemia in offspring: the Ontario case-control study', *BMJ* Vol. 307 (6910); 16 Oct. 1993.

Mitchell, P. and Vintró, Luis Leon, Department of Experimental Physics, UCD, 'What is the environmental Impact of Sellafield discharges?' (abstract), RIA 'Making Sense of Sellafield' Conference, 2003.

Murnaghan, D.J. 'History of radium therapy in Ireland', *Journal of the Irish Colleges of Physicians and Surgeons*, Vol. 17, No. 4, October 1988.

Murray Patrick, 'Obsessive historian: Eamon De Valera and the policing of his reputation', Royal Irish Academy, Dec. 2001.

Smith, P.G. and Douglas, A.J., 'Mortality of workers at the Sellafield plant of British Nuclear Fuels', *BMJ* Clin. Res. Ed., Vol. 293, No. 6551, 1986.

Tweena, Michelle, 'Nuclear Energy – Rise, Fall and Resurrection', Center for International Climate and Environmental Research, March 2006.

Archives, Reports and Records

'Chernobyl's Legacy: Health, Environmental and Socio-Economic Impacts,' Chernobyl Forum, UN, September 2005.

ESB Reports, Carnsore Project

National Archives, Kew UKAEA Files

NEB Files, RPII Library, National Library of Ireland

Oireacthas Éireann: Dáil and Seanad Official Reports and Transcripts 1922–2007

RPII Annual Reports, Reports on Sellafield

UNSCEAR 1993 Report

Newspapers and Media (Ireland and UK) 1927–2007

Cork (Irish) Examiner
Daily Mirror
Daily Telegraph
Guardian
Independent (UK)
Ireland on Sunday

Irish Independent
Irish Mirror
Irish Press
Irish Times
New Scientist
Sunday Independent
Sunday Press
Star
Sunday Times (Ireland Edition)
Sunday Times (UK)
Whitehaven News & Star

98FM
LMFM
RTE (TV & Radio News, Current Affairs Programmes)

Index

Acheson Lilienthal report 75
Acheson, Dean 78
Adam, Gerry 216–7
Ahern, Bertie 204, 221, 232, 237, 238, 239, 246, 249, 253, 257, 265, 267
Ahern, Dermot 231–2
Ahern, Nuala 236
Acquired Immuno-Deficiency Sybndrome (AIDS) 221
Aiken, Frank 52, 122, 123, 124 154–5
Al Qaeda 253
Allier, Lt. Jacques 36–7
Americium 126, 180, 230
Anderson, Frank 108
Anderson, Sir John 65
Anglo-Irish Agreement, 1985 189
Anglo-Irish Trade Agreement, 1938 27
Anglo-Irish Treaty 3, 4, 21, 23, 24, 27, 28, 203
Askew, Norman 261, 262
Atlee, Clement 43, 46, 53, 61–3, 64–5, 67, 70, 76, 76, 78, 92, 93, 94
Atomic Energy Committee (AEC) 84–5
Aziz, Tariq 161

Barnaby, Dr. Frank 150, 252
Barry, Peter 159
Baruch, Bernard 76
Battle, John 246
Beckett, Margaret 249, 261
Becquerel, Henri 8, 9
Bell, Michael 233
Benn, Tony 134, 135, 141, 143, 144, 180
Berry, Dr. Roger 211–2
Beryllium 18
Bethe, Hans 69–70, 81
Bevin, Ernest 46, 62, 65, 66, 76, 92, 93, 159
Biggs, Sir Christopher Ewart 136
Bin Laden, Osama 253
Black report 177, 178, 189–90, 191, 210
Black, Sir Douglas *see* Black report
Blair, Tony 237, 238, 242, 246, 253, 261, 264
'Bloody Sunday' 136
Bohr, Niels 11, 16, 29, 30, 31, 35, 37, 39, 42, 73
Bolter, Harold 141, 142–3, 144, 207, 212

Bonser, David 217–8
Born, Max 16
Boron 30
Bowler, John 150, 194–6, 202, 217, 234, 283
Brent Spar Campaign 234
Briggs Committee 39
British Medical Journal 211
British Nuclear Fuels Ltd.(BNFL) xiii, 10, 127, 130–3, 135, 139, 140–2, 143, 144, 145, 147, 148, 163, 175, 176, 179, 180–1, 182, 206, 208, 209, 210, 211, 212, 213–7, 224. 226–31, 232, 233, 234, 235, 237, 239, 240–6, 247, 249, 250, 251, 252, 255–6, 257, 260, 261–2, 264, 265, 267, 272–3, 280–1, 282–3
British Nuclear Group (BNG) 261–2
Browne, Dr. Noel 116, 117, 123
Bruton, John 165–6, 219, 223–4, 237
Bruton, Richard 201–2
Bryers, Stephen 246
Burke, Raphael P. (Ray) 165–6, 182, 195, 198, 199, 200, 201, 202–3, 220
Bush, Vannevar 45
Byrnes, Martin 66

Caesium 146, 147, 179, 186
Callaghan, Jim 136–7, 144, 158
Campaign for Nuclear Disarmament (CND) 167, 194, 232
Carbon dating 10
Carter, Gavin 217, 243
Carter, Jimmy 133, 156
Cavendish Laboratory, Cambridge 7, 8, 13–14, 15–21, 45, 53
Central Electricity Generating Board (CEGB) 197, 201, 208
Central Intelligence Agency (CIA) 245, 253
Chadwick, James 12, 18, 20, 41, 42, 47, 53, 64, 67, 73–4, 275
Chamberlain, Neville 27, 28, 42–3, 50, 212
Chernobyl accident xii, 96, 172, 184, 185–9, 190, 197–8, 230, 239, 254, 255, 264, 278, 280
Cherwell, Lord 45, 74, 80–1, 88–9, 93–4

Churchill, Winston 28, 42, 45, 47, 48–9, 51–2, 61, 62–3, 65, 73, 75, 76, 80, 82, 87, 88, 89, 94, 276
Clarke, John 256
Climate change xi–xii, xiii
Coakley, Sean 151
Cockcroft, John 17–18, 19–21, 42, 44, 47, 48, 64, 66, 67, 69, 72, 77, 91, 93, 94–5, 98, 107, 108, 128, 274–5
Cocker, Prof. Wesley 71–2
Cold War 71, 88, 123, 228, 274
Colley, George 155, 167, 168
Collum, Hugh 241, 261
Colville, Sir John 82
Comber Dr. Harry 236
Combined Development Agency (CDA) 79–80
Comhairle, Ceann 194
Committee on Medical Aspects of Radiation in the Environment (COMARE) 178, 213
Conant, James 45
Conroy, John 84
Cork Anti-Nuclear Alliance (CANA) 154
Cosgrave, Liam 83, 84, 85, 121, 122, 139, 205
Cosgrave, William T. 4, 21, 24, 26, 122
Costello, Declan 122
Costello, John A. 70–1, 83, 84, 116, 117
Coulston, David 244
Cowen, Brian 210, 217, 222, 227, 237
Cox, Senator Margaret 156
Coyle, Rupert 164
Craigavon, Lord 51
Cripps, Stafford 65, 66
Cuban missile crisis 123
Cullen, Martin 266
Cumann na nGaedheal 4–5, 22–3, 24–5, 276
Cunningham, Chris 152
Cunningham, Jack 143, 180, 206, 207
Cunningham, John 164, 184, 188
Curie, Marie 3, 8, 9, 11, 12, 16, 42
Curie, Pierre 9, 11, 12, 16
Currency Act, 1921 25
Cutler, James 175
 'Windscale–The Nuclear Laundry' 175–6, 177, 182

Dáil Éirann 4, 21, 27, 48–9, 51, 71, 85, 111, 115, 121, 122, 138, 155, 163, 182, 188–9, 193, 194, 195, 203, 218–9, 220, 231, 237, 246, 249, 250
Daily Herald 108
Daily Mail 21
Daily Mirror 128, 135

Daily Telegraph 90, 93, 261
Dalton, Hugh 65, 66
Davey, Hugh Gethin 61, 69, 96, 103, 104, 108, 111, 171
Davis, Stanley Clinton 190, 197, 199
Day, Martyn 212
De Valera, Eamon 3–4, 21–2, 23, 25–8, 58–52, 53, 70, 85, 111, 116, 117, 122, 164, 167, 203, 276
Dean, Dr. Geoffrey 237, 247
Dempsey, Renee 269
Desmond, Barry 137–8
Dewhurst, Philip 261
Dillon, James 71
Dixon, Daniel *see* Glentoran, Second Lord of Antrim
Doll, Richard 117–8, 211
Downing St. Declaration 219–20
Doyle, Avril 232–3
Duffy, Dr. George 198, 199
Duncan, Colin 243, 245, 246, 250
Dunkirk retreat 43
Dunster, John 96–7, 105

'Economic War' 26
Economist 109
Einstein, Albert 10, 11, 12, 16, 20, 29, 30, 31, 37–8, 39, 123, 272
 Special theory of relativity 10, 20, 30, 272
 General theory of relativity 16
Eisenhower, President Dwight D. 82–3, 88, 89, 100–1
Electricity Regulation Act, 1999 240
Electricity Supply Board (ESB) 137–9, 149, 151–3, 155, 158–9, 160, 162, 163, 164, 165, 198, 277
Euratom Treaty 125, 183, 190, 195, 210
Extradition Act, 1986 194, 204

Fahy, Frank 154
Falklands War 182
'Fat Man' 54–5
Fermi, Enrico 29, 30, 35, 37, 39, 46, 48, 73
Fermi, Laura 29
Fianna Fáil 4, 21–2, 23, 25, 26, 50, 85, 111, 116, 117, 158–9, 166, 182, 188, 190, 192–4, 195, 202, 203, 204, 217, 218–9, 223, 231, 232, 237, 240, 247, 250, 257, 259, 276
Financial Times 141
Fine Gael 70, 116, 118, 122, 138, 139, 149, 159, 162, 165, 166, 168, 182, 192, 193, 201, 202, 203, 204, 219, 223, 232, 233, 252, 265, 282

Finlay, Fergus 263
Finucané, Marian 253–4
Firth, George 164
FitzGerald, Garret 136–7, 166, 181–2, 189, 191–2, 193–4, 204, 277–8
Flerov, Georgi 39–40
Flynn, Padraig 195–6
Ford, Gerald 133
Fox, Mildred 246
Free State Executive Council 21
Free Trade Treaty, 1965 117
Freemasonry 245
Friends of the Earth 135, 141–2, 143, 162, 176, 195, 214, 216, 249, 250, 280, 281
Frisch, Otto 30, 40–1, 42
Fuchs, Klaus 74–5, 78

Gamma rays 18, 99
Gamow, George 18, 19, 20
Gardaí 23, 26
Gardner, Prof. Martin 210
Geiger, Hans 17
Gen. 163 66
Gen. 75 65–6
Ghaddafi, Col. 134
Gilmore, Eamon 232
Glentoran, Second Lord of Antrim 118–22, 135, 276
Gold, Harry 75
Good Friday Agreement 204, 238
Goodlet, B.L. 87
Government of Ireland Act, 1920 3
Grannard, Lord 21
Great Depression 23, 25
Great Exhibition 7
Great War 4, 5, 12–13, 17, 28
Greenpeace 135, 149, 150, 176, 179–80, 181, 192, 194, 195, 196, 197, 202, 207, 213–4, 215, 217–8, 220, 221, 222, 223–6, 245, 249, 250, 252, 264–5, 280–1, 282, 283
Grehan, Dr. Mary 235–6
Groves, Gen. Leslie 46, 47, 65, 66, 67, 74, 91
Guardian 103, 179–80
Guinness, John 227–8, 229, 232, 241
Gummer, John Selwyn 215–6, 219–20, 223, 224, 229
Gwynn, E.J. 23

Hahn, Otto 30
Halifax, Lord 43
Hamilton, James 225
Hammarskjold, Dag 82
Harney, Mary 222

Harrisburg accident 156–8, 164, 165, 168, 172–4, 277, 280
HASS Directive 268–9, 284
Haughey, Charles J. 167–8, 171, 181, 188–90, 191, 192, 194, 195, 198, 201, 203–5, 219, 277
Healey, Denis 137
Hearne, John 84
Heath, Edward 127, 134
Heisenberg, Werner 16
Hempel, Carl G. 53
Hesketh, Ross 125
Hewson, Ali 263, 264
Hill, Sir John 140, 147
Hillery, Dr. Irene 182
Hinton, Sir Christopher 66, 67, 68, 86, 87, 91, 94–6, 99, 275
Hiroshima 54–5, 61, 70, 91, 110, 123, 149, 161, 211, 273
Hitler, Adolf 13, 28, 29, 36, 39, 43, 272–3, 274
Holocaust 42
Hope & Reay Case 212–3
Howlin, Brendan 232, 237
Hughes, Emrys 93
Hurd, Douglas 280
Hussein, Saddam 161
Hyde Park Agreement 63

Independent 242–5, 261, 262
Innes, Jim 205, 206–7, 211, 214, 215, 218, 275
International Atomic Energy Agency (IAEA) 82, 155, 190
International Commission for Radiological Protection (ICRP) 94, 105, 146, 173, 258
Iodine 131 105–6, 172, 174, 178, 186, 188, 255
Ireland on Sunday 253, 263
Irish Atomic Energy Commission (IAEC) 122, 241
Irish Civil War 3, 4, 6, 21, 116, 117
Irish Independent 21, 252
Irish Press 21, 160, 166
Irish Republican Army (IRA) 3, 4, 22, 23, 48–9, 71, 120, 136, 168, 182, 204, 219, 223
Irish Times 21, 52, 83, 84, 90, 107, 110–11, 234, 246, 284
Irish War of Independence 3, 4, 65

Jacob, Joe 238, 240, 246–8, 249, 253
Jenkin, Patrick 191
Jenkins, Les 109–10
Johnson, Alan 261

Joint Committee on Atomic Energy (JCAE) 64, 100
Joliot Curie, Frederic 18, 29–30, 35, 36–7, 39, 40
Joliot Curie, Irene 18, 29–30, 37
Joly, John 12, 14
Jones, John Paul 60
Jordan, Pascual 16

Kelley, John 149
Kelly, Petra 153
Keynes, John Maynard 61–2
Kinlen, Leo 178
Korean War 78
Kowarski, Lew 35, 37

Large, John 150, 199, 220, 252
Lauterpacht, Elihu 225
Lawler, Liam 217
Lawrence, Ernest Orlando 15, 19, 47, 73
Lawson, Nigel 207
League of Nations 24, 50
Lemass, Sean 26, 115, 117–8, 137, 167
Lenihan, Brian 138–9
Leslie, Frank 108
Liddell, Helen 244
Lindemann, Sir Frederick *see* Cherwell, Lord
Lithium 20
'Little Boy' 54–5
'Little Joe' 69, 77
Little, Arthur D. 248
Liverpool Daily Post 108
London Evening Standard 261
Loughlin, Chris 243
Louth Four 235–7, 247, 267
Lowry, Dr. Sidney 191
Lowther, Sir John (Earl of Lonsdale) 59
Lynch, Jack 166–7

MacDonald, Malcolm 50–1
MacLean, David 214
MacMillan, Harold 83, 100–1, 105, 106, 107, 109, 111, 112
Major, John 219, 223–4
Makins, Roger 78
Manchester Literary and Philosophical Society 10–11
Manhattan Project 15, 46–8, 54, 70, 73, 79, 91, 123, 273
Mansergh, Martin 167–8, 220, 238
Marconi, Guglielmo 7
Marshall Aid Fund 71, 76
Marshall Plan 76–7, 115
MAUD Committee 42, 44–5
Maud, Sir John 120

May, Alan Nunn 63–4
Mayak explosion 147
McAulay, Dr. Ian 200
McBride, Séan 71
McCartan, Pat 202, 203
McCarthy, Joe 78
McDowell, Michael 246, 249, 258, 265–6
McGarry, Ann 269
McKeown, J. A. 120
McLaughlan, Tom 216, 243
McLaughlin, Dr. P.J. 110–11
McLean, Andrew 105
McLean, Donald 75, 77
McMahon Act, 1946 64, 91, 100
Medical Research Council (MRC) 99
Medvedev, Zhores 146–7
Meitner, Lise 30, 40, 42
Mitchell, Dr. Peter 188
Modus Vivendi 77, 79–80, 81
Moore, Jo 246
Morgan, Arthur 236
Moseley, Harry 11, 12
Muggeridge, Malcolm 93
Mulcahy, Richard 83
Mummery, Peter 140, 141–2, 175–6, 211
Mussolini, Benito 29, 36
Mutual Defence Agreement (MDA), 1958 125

Nagasaki 54–5, 61, 66, 110, 149, 161, 211, 241, 273
National Health Service 62
National Radiological Protection Board (NRPB) 174, 177
National Union of Miners (NUM) 205
Nature 18, 20, 36
National Campaign fir the Nuclear Industry (NCNI) 206, 207, 214–5, 218, 244, 270
Ne temere decree, 1904
New Scientist 111, 156
Newsweek 187
Non-Proliferation Treaty, 1967 123
North Atlantic Treaty Organisation (NATO) 71, 78, 82, 100, 276
Nowlan, Noel 164, 198
National Radiological Protection Board (NRPB) 202
Nuclear Energy Board (NEB) 138, 149, 152–3, 154, 164, 165, 182, 183–4, 187–8, 190, 195, 196, 197–200, 203
Nuclear Installations Inspectorate (NII) 144, 197, 198–9, 242, 244
Nuclear waste xi, 68, 93, 95–7, 130–1, 144, 145, 147, 162, 179–81, 183, 203, 207, 223, 224–6, 230, 244, 282

O'Donnell, Tom 138
O'Dowd, Fergus 233, 236, 238, 247, 252, 255, 256–7, 267
O'Duffy, Eion 23, 26
O'Farrelly, Christopher 151, 158
O'Flaherty, Dr. Tom 254–5
O'Higgins, T. F. 118
O'Malley, Des 159–64, 165, 167, 168, 204, 219, 277
O'Neill, Capt. Terrence 118
O'Rourke, Mary 240
Oliphant, Marc 40, 42, 47, 48, 53, 274
Oppenheimer, Robert 15, 31, 48, 73, 75
Organisation for Eurpoean Economic Co-Operation (OEEC) 76, 82–3, 84

Paisley, Ian 257
Parker, Justice Roger 143
Parkinson, Cecil 199, 208
Patterson, Walt 135, 143
Peierls, Rudolf 40–1, 42
Penney, Sir William 66, 93, 101, 109, 201, 275
Perrott, Sir Donald 120
Physical Review 37
'Pippa' 87–8
Plowden, Sir Edwin 105, 107, 112, 119, 120, 121
Plutonium xii, 37, 46, 54, 61, 62, 64, 66, 67, 69, 77, 78, 79, 81, 87, 88, 89, 90, 96, 98, 99, 101, 102, 110, 124, 125, 126, 130, 131, 133, 145, 146, 147, 154–5, 174, 180, 185, 20, 230, 240–1, 251, 265, 270–1
    Plutonium 239 ('weapons-grade plutonium') 88, 90, 125, 126, 130, 280
    Plutonium 240 ('civil-grade plutonium') 88, 125
Polanyi, Michael 11
Pollard, David 255
Polonium 9, 102
    Polonium 210 102, 174
Portal, Lord 56, 67–8, 77, 87
Potsdam Conference 54, 55
Prince Albert 7
Prince Charles 263, 264
Public Order Act 23
Pugwash movement 123

Quebec Agreement 47, 63, 64, 79
Queen Elizabeth II 86, 90, 239
Queen Victoria 7
Quinn, Ruairi 155, 250

Radio Tefefís Éirann (RTE) 52–3, 263
Radiological Protection Institute of Ireland (RPII) 183, 227, 231, 232, 244, 247, 254, 255, 256, 258, 264, 267–8, 282
Radium 9, 11, 12
Radon 9, 236
Rainbow coaltion 223–4, 232, 237
Rasmussen report 150
Rees, Merlyn 137
Reville, Dr. William 154–5
Reynolds, Albert 219
Ritson, Stan 108
Roache, Adi 263–4
Roache, Dick 267, 296
Roentgen, Wilhelm 7, 8, 11
Roosevelt, Franklin D. 38, 39, 45–6, 47, 51, 53, 54, 63, 73
Rosenberg, Ethel 75
Rosenberg, Julius 75
Ross, K.B. 105
Rotblat, Joseph 41–2, 73–4, 92–3, 123, 149
Rotblat, Tola 41–2
Royal Society, London 15, 21
Russell, Bertrand 123, 167
Ruthenium 96–7
Rutherford, Ernest 7, 8, 9–11, 12, 13–15, 17–18, 19, 20–1, 24, 29, 31, 272

Sachs, Alexander 38
Salisbury, Lord 119, 120
Sargent, Trevor 220–1, 240
Scargill, Arthur 205–6
Schneider, Mycle 251–254
Schonland, Dr. B.J.F. 109
Schrödinger, Erwin 16
Sea Shepherd 135, 144
Seaborg, Glenn 45
Second World War xiii, 28–9, 38–59, 61, 62, 73, 79, 100, 120, 171, 194, 272–4, 276, 279
September 11 attacks 249–52, 253–7
Sheehan, Dr. Patricia 182–3, 191, 235, 237, 278
Shore, Peter 142–3, 177
Simon, Franz 44
Sinn Féin 3, 4, 25, 115, 118, 204, 216, 236, 267
Sinton, Anna 5, 6
Smith, Michael 202, 203, 254
Smith. Dr. Peter 212
Smylie, Robert 52
Smyth report 91
Snow, C.P. 53, 274
Soddy, Frederick 9–10
Solvay Conference, Fifth International 16

Spanish Civil War 29
Spring, Dick 187, 190–1, 193, 198, 218–9, 263
Sputnik I 101
Stagg, Emmet 223–5, 226, 230–1, 232, 234, 239, 240, 249–50, 264
Stalin, Josef 39, 54, 70, 76, 82
Star 253, 264
Steevens, Dr. Walter 12
Stokes, Richard 96
Strassman, Fritz 30
Strauss, Lewis 86, 94, 147
Strontium 147
    Strontium 89 99
    Strontium 90 99, 221
Suez crisis 90, 115, 129
Sun 216–7
Sunday Tribune 231
Sunningdale Executive 136
Sutherland, Veronica 226
Swift. Jonathan 59
    Gulliver's Travels 59
Szilard, Leo 29, 30–1, 35, 37–9, 47, 48, 55, 97–8, 273

Taylor, John 228, 242, 243, 244
Taylor, Peter 174
Technetium 230, 266
Teller, Edward 29, 37, 97, 98, 102, 151, 273–4
Tellurium 186
Tenet, George 253
Thatcher, Margaret 144, 176–7, 181–2, 189, 191, 204, 205, 277
Thompson, Dr. Gordon 231
Thomson, J. J. 8, 10, 13
Thomson, Sir George 42
Thorburn, Richard L. 157
Thorium 9, 11
Three Mile Island accident *see* Harrisburg accident
Thrift, Prof. William E. 7
Times 90, 118, 257, 259
Tizard, Sir Henry 42, 66–7
Tritium 102, 125, 174, 217
Truman, President Harry S. 54, 55, 62, 63, 76, 77, 78
Tuohy, Tom 104–5
Tuvey, Frank 164, 184, 197–200, 201, 231, 244

U2 214, 263
UK Atomic Energy Authority (UKAEA) 89, 95, 96, 97, 99, 105, 106, 107, 108, 109, 111, 112, 119, 120, 121, 127, 130, 140, 176, 187, 240, 250, 276, 281
Uranium 9, 30, 35, 36, 37, 38, 44, 47, 54, 63, 68, 77, 78, 79, 80, 81–3, 84, 86, 90, 95, 96, 98, 99, 102, 103, 109, 120, 121, 125, 127, 130, 131, 133, 154–5, 174, 185, 190
Uranium 235 37, 40, 42
Uranium 238 37
US Atomic Energy Act, 1946 64
US Atomic Energy Council (USAEC) 64, 86, 94, 100, 107

Von Halban, Hans 35, 37

Walker, Peter 195
Walton, Philip 274, 277
Walton, Prof. Ernest Thomas Sinton 3, 5–7, 13–15, 16–17, 18, 19–21, 23, 24, 25, 52–3, 71–2, 84–5, 123, 124, 274
Washington Declaration 63, 75
Wells, H. G. 31, 87, 273
    The World Set Free 31
Wertenstein, Ludwik 42
Wesley, John 5
Wheeler, John 37
Whitaker, T.K. 117
Whitehaven News 108, 111–12
Wigner, Eugene 29, 37, 39, 46, 98
Wilcox Baker, Rupert 227, 231, 233, 246, 256
Williams, Lawrence 242
Wilson, Arthur 103–4
Wilson, Brian 238, 248, 258–60, 279, 280
Wilson, Freda 6, 19, 20–1, 24
Wilson, Harold 127, 134, 135, 137
Windscale fire xii, 96, 101–12, 147, 163, 172, 174, 182, 186, 196, 187, 201, 235, 247, 270, 278, 285
Woods, Martha 6
Wren, Sir Christopher 59

X-rays 7–8, 12, 17, 38, 53

Yeats, W. B. 25